T0240014

COVID-ology

The COVID-19 outbreak and response have been characterized by institutional missteps, media misinformation, and economic and social upheavals. This book is intended to de-mystify and inform in a succinct and straightforward text. The lessons shared are applicable to other infectious diseases and future inevitable pandemics. Readers will learn about the origins of SARS-CoV-2, the disease it causes, clinical tests and how they work, therapeutics, and prophylactic measures such as vaccines. This information will prepare readers to be better able to respond to future emerging infectious diseases and pandemics.

Key Features

- Positively influences career choices within public health
- Applies basic science to problems raised by the COVID-19 pandemic such as vaccine development and herd immunity
- Prepares readers with context and tactics for understanding future infectious disease outbreaks
- Successfully used in college senior health sciences seminars
- Engaging and balanced treatment of the politicization of public health issues, especially COVID-19

COVID-ology

A Field Guide

Dr. Michael T. Myers, Jr., MD, MBA

CRC Press
Taylor & Francis Group
Boca Raton London New York

CRC Press is an imprint of the
Taylor & Francis Group, an **informa** business

Cover Image Credit: Shutterstock.com

First edition published 2023
by CRC Press
6000 Broken Sound Parkway NW, Suite 300, Boca Raton, FL 33487-2742

and by CRC Press
4 Park Square, Milton Park, Abingdon, Oxon, OX14 4RN

CRC Press is an imprint of Taylor & Francis Group, LLC

ISBN: 9781032316055 (hbk)
ISBN: 9781032316031 (pbk)
ISBN: 9781003310525 (ebk)

DOI: 10.1201/9781003310525

Typeset in Minion
by Deanta Global Publishing Services, Chennai, India

To all future public health leaders

Contents

Acknowledgments

Many heartfelt thanks to Tara Blaine, extraordinary collaborator, writing coach, and editor, without whose recommendations and reassurance this field guide would not have been possible, and to Jason McCann for introducing me to Tara. To Karen Kovach for indexing the work and copyright clearing all the images. To Stonehill College, especially Father John Denning, Heather Heerman, and Katharine Harris, who gave me the chance to teach "Topics in Health Science: *COVIDOLOGY*" to six of its brightest students: Destinee Alix-Garth, Courtney Marcos, Megan MacIver, Emily O'Brien, Kristen Thornton, and Bridget Wetmore. To Ben Kerman for Sunday morning chats about the ideas of the book and to Jeff Bloch for all-day-or-night mentoring, encouragement, and introducing me to Tim Stookesberry, who convinced me to publish with CRC Press. To Chuck Crumly at CRC Press who took immediate interest in the project and to Kara Roberts for shepherding the work through editorial production. To my sister Mauri Myers-Solages for photography, laughter, and the real real. To my husband Bob for hanging in there as the book was written during a January through April, 2022 transcontinental trek across America. And to my mother, Joyce Ann Williams Myers, the original Tiger Mom, who made sure I had the best education money couldn't buy when she convinced the Kansas City Missouri School Board in 1971 that a black kid living on the wrong side of Troost Avenue should be admitted to the finest schools in the district, George Caleb Bingham Junior High School and Southwest High School, which are now sadly demolished or closed.

Author

Michael T. Myers, Jr., MD, MBA is an Adjunct Lecturer at Stonehill College in Easton, Massachusetts, where he teaches about COVID-19 and is scientific advisor to its COVID-19 Testing Program. With 33 years of experience in health care, Dr. Myers is Chief Medical Officer of Compass Medical, among the largest primary care groups in Massachusetts. Dr. Myers earned his MD from Harvard Medical School, MBA from Northeastern University, and BA from Johns Hopkins University.

Acronyms

ACE2	angiotensin converting enzyme 2
ACIP	Advisory Committee for Immunization Practices
ACT-A	Access to COVID-19 Tools Accelerator
ADME	absorption, distribution, metabolism, and excretion
AEC1	Type I alveolar epithelial cells
AEC2	Type II alveolar epithelial cells
AIDS	acquired immunodeficiency syndrome
APA	Administrative Procedures Act
APC	antigen-presenting cell
ARDS	acute respiratory distress syndrome
ASPR	Assistant Secretary for Preparedness and Response
BSL	biosafety laboratory
BLA	Biologics License Application
CBRN	chemical, biological, radiological, and nuclear
CEPI	Coalition for Epidemic Preparedness Innovation
CONSORT	Consolidated Statement of Reporting Trials
COVAX	COVID-19 Vaccines Global Access Facility
CDC	Centers for Disease Control and Prevention
CMV	cytomegalovirus
CYP3A4	cytochrome P4503A4
DHHS	Department of Health and Human Services
DNA	deoxyribonucleic acid
EID	emerging infectious disease
EMA	European Medicines Agency
EPI	Expanded Programme on Immunization
EUA	emergency use authorization
FDA	Food and Drug Administration
FFDCA	Federal Food, Drug and Cosmetic Act of 1938
HBV	hepatitis B virus
HCV	hepatitis C virus
HICs	high-income countries
HIV	human immunodeficiency virus
HPV	human papillomavirus

IAVG	Independent Allocation of Vaccines Group
ICC	Interstate Commerce Commission
ICTV	International Committee on Taxonomy of Viruses
IND	Investigational New Drug
IRB	Institutional Review Board
LFIA	lateral flow immunoassay
LMICs	low- and middle-income countries
MAB	monoclonal antibody
MCM	medical countermeasures
MERS	Middle East respiratory syndrome
MHC	major histocompatibility complexes
MHRA	Medicines and Health Products Regulatory Agency
MIS-C	multisystem inflammatory syndrome in children
MMR	measles, mumps rubella vaccine
MMWR	Morbidity and Mortality Weekly Report
mRNA	messenger RNA
NDA	new drug application
NIH	National Institutes of Health
NPI	non-pharmaceutical interventions
NPV	negative predictive value
NSP	nonstructural protein
NTD	neglected tropical disease
OCP	Onchocerciasis Control Programme
ORF	open reading frame
PANGO	Phylogenetic Assignment of Named Global Outbreak lineage
PASC	post-acute sequelae of COVID-19
PCR	quantitative real-time polymerase chain reaction test
PHSA	Public Health Service Act of 1944
PPV	positive predictive value
R&D	research and development
RAAS	renin-angiotensin-aldosterone system
RCT	randomized clinical trial
RDT	antigen rapid diagnostic test
RdRp	RNA-dependent RNA polymerase
RNA	ribonucleic acid
rRNA	ribosomal RNA
RSV	respiratory syncytial virus

SARS	severe acute respiratory syndrome
SARS-CoV-2	severe acute respiratory syndrome coronavirus 2
TMPRSS2	transmembrane protease serine S2
tRNA	transfer RNA
UNICEF	United Nations Children's Fund
VRBPAC	Vaccines and Related Biological Products Advisory Committee
VZV	varicella-zoster virus
WHO	World Health Organization
WIV	Wuhan Institute of Virology

COVID-19 Timeline

COVID-19 TIMELINE

December 12, 2019 Small number of pneumonia cases of unknown cause are seen in Wuhan, Hubei Province, China

January 3, 2020 44 cases of pneumonia of unknown origin in Wuhan, China, are reported to WHO

January 10, 2020 China releases first draft of genome sequence for 2019 novel coronavirus

January 20, 2020 CDC confirms first US case of 2019 novel coronavirus in Washington state

January 30, 2020 WHO declares SARS-CoV-2 outbreak a Public Health Emergency of International Concern

January 31, 2020 Alex Azar, US Secretary of Health and Human Services, declares SARS-CoV-2 a Public Health Emergency, with new travel bans effective February 2, 2020

February 3, 2020 WIV publishes first complete genomic sequencing of SARS-CoV-2 virus

February 11, 2020 International Committee on Taxonomy of Viruses (ICTV) announces "severe acute respiratory syndrome coronavirus 2 (SARS-CoV-2)" as the name for the 2019 novel coronavirus

WHO designates the abbreviation of coronavirus disease 2019, "COVID-19", as the official name for the outbreak caused by the 2019 novel coronavirus, SARS-CoV-2

February 23, 2020 Italy institutes emergency orders to lock down the country in light of dramatic rise in cases

March 11, 2020 WHO declares COVID-19 a global pandemic

March 13, 2020 President Trump declares a national emergency in the US

March 16, 2020 Moderna begins phase 1 trial of mRNA vaccine (mRNA-1273)

April 21, 2020   NIH posts first set of "NIH COVID-19 Treatment Guidelines"

April 23, 2020   President Trump suggests investigation into internal ingestion of disinfectants as a COVID-19 treatment

May 2020   Sweden announces results from "natural immunity" social experiment

May 5, 2020   Pfizer/BioNTech enter phase 1/2 trials of mRNA vaccine (BNT162)

May 15, 2020   Operation Warp Speed launches

October 2, 2020   President Trump tests positive for SARS-CoV-2

October 4, 2020   Great Barrington Declaration

October 14, 2020   John Snow Memorandum

November 9, 2020   Pfizer/BioNTech announce BNT162 has efficacy greater than 90%

November 16, 2020   Moderna announces mRNA-1273 has efficacy of 94%

December 8, 2020   Margaret Keenan of the UK becomes first person in the world to receive COVID-19 vaccine

December 11, 2020   FDA grants EUA to Pfizer/BioNTech for use of BNT162

December 18, 2020   FDA grants EUA to Moderna for use of mRNA-1273

July, 2021   Delta becomes predominant circulating SARS-CoV-2 variant

July, 2021   Surgeon General issues Advisory warning against the dangers of misinformation

November, 2021   Omicron becomes predominant circulating SARS-CoV-2 variant

March, 2022   10 WHO-authorized vaccines are in use worldwide

May 17, 2022   United States surpasses one million COVID-19 deaths

Introduction: Foray into the Field

COVID-19 UPDATE: MAY 9, 2022

On May 9, 2022, a little over two years since the World Health Organization (WHO) declared COVID-19 to be a global pandemic,[1] there were 518,058,774 recorded global cases of the disease.[2] Of those, 16%—a total of 82 million—occurred in the United States, which holds only 4% of the world's population. The global tally of deaths from COVID-19 stood at 6.3 million; nearly 1 million of those deaths were in the US, 16% of the global total.

Of the five variants of the SARS-CoV-2 virus,[3] Omicron was demonstrating its ability to surpass its predecessors in terms of transmissibility. By May 2022, the most circulating SARS-CoV-2 virus was the BA.2 subvariant, with BA.2.12.1 beginning to rise, and concerns were growing about the new strains BA.4 and BA.5. SARS-CoV-2 had not yet mutated to a form warranting a new Greek letter designation.

Vaccination rates among US residents were robust, with 83% having completed at least one dose of three available COVID-19 vaccines authorized for emergency use by the Food and Drug Administration (FDA).[4] But other countries, especially those on the African continent, were faring less well. Nigeria—Africa's largest nation, with over 200 million people—had only managed to vaccinate 13% of its residents. South Africa also had a suboptimal vaccination rate of 36% and was thought to be the source of the BA.4 and BA.5 subvariants related to the Omicron strain of SARS-CoV-2.

In addition to several safe and effective vaccines for worldwide use, there were a number of new antiviral medications—including Paxlovid™,

DOI: 10.1201/9781003310525-1

1

remdesivir, and monoclonal antibodies—which could shorten the disease course of COVID-19, making it far less severe.

A Bewildering Landscape

SARS-CoV-2 took the world by surprise. Three serious infectious disease outbreaks earlier in the twenty-first century—SARS, MERS, and Zika—did not affect enough people, cause substantial deaths, or last long enough to capture the world's attention.

But COVID-19 was an arresting, once-in-a-lifetime experience. Prior to the identification of the SARS-CoV-2 virus as the causative agent for the disease, it was referred to as "novel coronavirus," an apt moniker for an unprecedented global public health crisis.

As the illness spread across the globe, it paralyzed world economies and challenged governments and public health leaders to rapidly develop response strategies. COVID-19, as an infectious disease phenomenon and public health emergency, became weaponized for political gain as facemask wearing and isolation curtailed individual liberties, and social media platforms allowed conspiracy theories and misinformation to metastasize. Awash in a sea of frightening facts and conflicting information, every individual was thrust into a new landscape fraught with unanswered questions: what exactly was this virus? How could we keep ourselves safe? When would it end—would it, in fact, ever end?

An Emerging Infectious Diseases Field Guide

The twenty-first century is challenged by three existential crises: thermonuclear war, global climate change, and emerging infectious diseases (EIDs).

EIDs have been a prominent feature of the last 20 years, with six significant EID outbreaks: SARS (2002), H1N1 influenza (2009), MERS (2012), Ebola (2014), Zika (2015), and the COVID-19 pandemic, which continues now. This new century has given birth to a threatening new world in which pandemics, like heat waves and hurricanes, will occur with greater frequency and ferocity.

COVID-ology: A Field Guide is a textbook for curious people to learn about the new world of EIDs. The COVID-19 pandemic has provided a wide range of issues to explore: what exactly are viruses, and how do they cause disease? Where did SARS-CoV-2 come from, and how will other pandemic-potential pathogens emerge? What role do government agencies like the Centers for Disease Control (CDC) and FDA play in regulating medications and vaccines? How were COVID-19 vaccines able to be developed in 11 months when this process usually takes 10 years or more?

Field guides help people understand and appreciate the natural world around them. They enable forays into new territory, equip explorers with knowledge and insight to understand what they see, and empower inquiring minds to unearth new discoveries. COVID-19 has shown that pandemic-level EIDs are an undeniable part of the natural world; studying this disease provides an opportunity to map the contours of the new landscape of EIDs, now and into the future.

Using This Field Guide

The first five chapters discuss the SARS-CoV-2 virus: how it functions, its origins, how it causes disease, clinical tests, and approaches to managing and treating COVID-19 illness. The last five chapters concern aspects of COVID-19 vaccines and attempts to control the spread of the virus: the role regulatory agencies played in COVID-19 vaccine development, important immunologic principles underlying all vaccines, herd immunity as a goal for vaccination programs, the enormous challenge of vaccinating the entire world, and human reactions to the pandemic that spurred the spread of infection. The conclusion looks ahead into the future of EIDs in the twenty-first century.

In each chapter, the reader is introduced to the topic with the following five elements:

- **The Field**: an overview of the curiosities that will be encountered
- **Field Sightings**: key terms and topics that will arise in the landscape
- **Field Expedition**: an initial outing into a case study related to the chapter's area of exploration
- **Lessons Learned**: notes on the import of the field expedition as it relates to COVID-19
- **Concepts Covered**: trail markers pointing to the main topics discussed

After delving into each chapter's topic, the reader will encounter opportunities for further exploration with three concluding elements:

- **Looking back, looking ahead**: reflections on the topics explored and pathways to future forays into the landscape of EIDs
- **Future Outings**: specific areas of inquiry required to further understanding of COVID-19 and future EIDs
- **Student Research Questions**: opportunities for readers to engage their curiosity and broaden their exploration of the chapter's key concerns

The Many Paths of Exploration

Throughout this field guide, readers will venture into multiple areas of scientific inquiry. These topics provide foundations for understanding the COVID-19 pandemic, as well as launching points for further study.

- **Genetics**: the study of genes, genetic variation, and heredity in living organisms
- **Virology**: the study of viruses and virus-like agents, including their structure, classification, and interaction with host organisms
- **Epidemiology**: the study of the causes, distribution, determinants, and control of diseases and health conditions
- **Pharmacology**: the study of the biochemical and physiological effects of chemicals that affect biochemical function in living organisms
- **Immunology**: the study of the characteristics and functioning of the immune system in living organisms

VENTURING INTO COVID-19—AND EIDS

This field guide equips students of the sciences and general readers to explore and understand the basic science and sociology of COVID-19. It can also serve as a complement to courses in molecular and cell biology, immunology, genetics, health care policy, and public health. The story of COVID-19 is ever-evolving, and its full implications will continue to reveal themselves as research continues.

Science, like any other human pursuit, is not carried out in a vacuum. It is intertwined with the fabric of society—with all its attendant human foibles and triumphs. Through a grounding in reliable scientific facts, *COVID-ology: A Field Guide* invites readers to discover the multifaceted story that is unfolding in this unique historical moment.

NOTES

1. This occurred March 10, 2020 when WHO officially called COVID-19 a " global pandemic," though its causative pathogen, a "novel coronavirus," had been circulating in China during the winter months of 2019.
2. Data retrieved October 5, 2022, 3:20 AM from COVID-19 Dashboard by the Center for Systems Science and Engineering (CSSE) at Johns Hopkins University (JHU). Dong E, Du H, Gardner L. An interactive web-based dashboard to track COVID-19 in real time. Lancet Infect Dis; published online February 19, 2020. https://doi.org/10.1016/S1473-3099(20)30120-1
3. Alpha, Beta, Gamma, Delta, and Omicron were the five "variants of concern" (VOCs) predominant during various phases of the first two years of the COVID-19 pandemic.
4. In May 2022, only people five years of age and older were eligible for COVID-19 vaccinations. Clinical trials for those younger than five years of age continued. The Centers for Disease Control and Prevention (CDC) defined "fully vaccinated" as having completed a primary series of vaccine: two doses of a messenger RNA vaccine from Moderna or Pfizer or one dose of the Johnson & Johnson adenoviral vector vaccine.

1

The SARS-CoV-2 Virus

The Field: What exactly is a virus, and how does it function? Unlike most organisms on Earth, a virus lacks the capacity to be self-sustaining—it cannot replicate itself. Instead, it sneaks into the cells of a host organism and takes over those cells' reproductive faculties. SARS-CoV-2 is a single-stranded positive sense RNA virus; its structure makes it particularly adept at commandeering host cells to replicate its genome. Understanding the specific mechanisms by which it does so is the foundation of all efforts to fight the disease this virus causes.

Field Sightings: astrobiology, ten-word definition of life, nucleotides, double helix, adenine (A), guanine (G), thymine (T), cytosine (C), codon, amino acids, uracil (U), translation, start codons, exons, introns, stop codons, Central Dogma of Molecular Biology, pathogen, genome (DNA or RNA), membranous envelope, organelles, ribosomes, endoplasmic reticulum, Golgi body (apparatus, complex), Baltimore system of virus classification, International Committee on Taxonomy of Viruses (ICTV), single-stranded positive sense RNA virus, open reading frame (ORF), 3CL-protease or main protease M^{pro} (NSP5), RNA-dependent RNA polymerase (RdRp; NSP12), angiotensin-converting enzyme 2 (ACE2), tropism, renin-angiotensin-aldosterone system (RAAS), cytoplasmic reproduction, virion, mutation, variant, lineage, Phylogenetic Assignment of Named Global Outbreak (PANGO) lineage, WHO Greek letter labels, S-gene target failure.

FIELD EXPEDITION: NASA'S TEN-WORD DEFINITION OF LIFE

Is there life on Mars? A serious attempt to answer this question occurred almost 50 years ago during the NASA Viking missions to Mars in 1976.

DOI: 10.1201/9781003310525-2

These missions successfully landed spacecraft on an extraterrestrial planet for the first time in human history.[1] Three biology experiments were conducted as part of the mission, and though no actual "life on Mars" was found, the mission ushered into being a new scientific discipline, **astrobiology**: "the study of the origin, evolution, distribution and future of life in the universe"[2] (see Figure 1.1).

As NASA began interplanetary exploration, it needed criteria for identifying living systems, should they be encountered. Based on what living organisms share on Earth, NASA proposed a **ten-word definition of life**: "Life is a self-sustaining chemical system capable of Darwinian evolution."[3]

FIGURE 1.1
Artistic depiction of Viking spacecraft over Mars releasing the lander descent capsule. (Source: Don Davis/Wikimedia Commons/Public Domain.)

The possibility of encountering forms of life beyond Earth—which may look radically different from those we know—required a clear definition of what, exactly, "life" is. Here on our home planet, it may seem like such a definition is obvious, but the tremendous variation among organisms muddies the waters. Viruses, in particular, confound the question of what, exactly, counts as a living thing.

Viruses are among the least complex of all organisms, consisting of only two parts: genetic material carried within a transporting envelope. This genetic material is either ribonucleic acid (RNA) or deoxyribonucleic acid (DNA) containing the coded information needed to make other viruses like itself. Would viruses satisfy NASA's ten-word definition of life? No and yes.

- *Life is a self-sustaining chemical system ...*: No, viruses are not self-sustaining; they require the internal machinery of host cells (especially ribosomes, endoplasmic reticula, and Golgi complexes) to replicate new versions of themselves.
- *... capable of Darwinian evolution.*: Yes, viruses mutate in response to environmental pressures, and this propensity demonstrates Darwinian evolution.[4]

LESSON LEARNED: NASA'S TEN-WORD DEFINITION OF LIFE

All viruses, including SARS-CoV-2, are "chemical systems" of genetic material carried within membranous envelopes. They cannot replicate new versions of themselves on their own, but they can usurp the machinery of host cells to make proteins needed for assembly into new viruses. While they fail the first half of NASA's ten-word definition of life, in that they are not "self-sustaining," their ability to take over a living organism's functions to replicate means that, in practice, they certainly do sustain themselves. And they are certainly capable of Darwinian evolution: all viruses have evolved to perpetuate themselves and adapt to changing environments through mutation, as notably demonstrated by the many variants of SARS-CoV-2.

Whether viruses are truly living systems according to NASA's ten-word definition is debatable, but when it comes to the human living organism, they are a distinct, separate entity. Understanding how viruses function is the cornerstone of fighting the diseases they cause. With a new viral emerging infections disease (EID), like SARS-CoV-2, scientists must identify the genome and life cycle specific to that virus—this is the foundation on which development of tools for fighting the disease within the human host is based.

CONCEPTS COVERED IN THIS CHAPTER

- Nucleic acids (DNA and RNA) as the self-sustaining chemical system of life
- The structure of viruses
- How viruses use host cells for replication
- Virus classification systems
- The SARS-CoV-2 genome and its replication process
- The SARS-CoV-2 life cycle
- Virus mutations, variants, and strains

Nucleic Acids Are the Self-Sustaining Chemical System of Life

NASA's ten-word definition of life as a "self-sustaining chemical system" is a direct reference to the chemical system upon which all terrestrial life depends: deoxyribonucleic acid (DNA) and ribonucleic acid (RNA). DNA contains the information needed to replicate an organism with complete fidelity, accomplishing this feat through a language based on chemicals called *nucleotides* (molecules consisting of a deoxyribose sugar, phosphate group, and nitrogenous base) and an organizational structure within a *double helix* (two parallel DNA strands bonded together and wound in a right-handed helix).

There are four nucleotides within all DNA molecules: the two purine nucleotides *adenine (A)* and *guanine (G)* and the two pyrimidine nucleotides *thymine (T)* and *cytosine (C)*. These four nucleotides are the genetic alphabet of life. Nucleotides proliferate one strand of a DNA molecule and are paired as purine-pyrimidine dyads to the opposing parallel strand: adenine (purine) is paired with thymine (pyrimidine), and guanine (purine) is paired with cytosine (pyrimidine).

A three-nucleotide configuration is called a *codon*, which encodes for a specific amino acid. *Amino acids* are the components of proteins, which have multiple functions in living systems. For example, the mRNA codon UGG codes for the amino acid tryptophan, which is represented by the one-letter abbreviation W (these one-letter abbreviations will be important in discussing viral mutations). There are 64 codons: 61 of them code for 20 amino acids, and the remaining three codons are "stop codons," discussed below. Amino acids, their three- and one-letter abbreviations, and corresponding codons appear in Table 1.1.

Ribonucleic acid (RNA) is the other important component of the self-sustaining chemical system upon which terrestrial life depends. RNA is comprised of four nucleotides containing a ribose sugar, phosphate group, and nitrogenous base. The two purine nucleotides in RNA are the same as those in DNA: adenine and guanine. RNA's pyrimidine nucleotides are cytosine (the same as DNA) and *uracil (U)* (which is not found in DNA). There are three forms of RNA, all of which serve different functions in protein synthesis: messenger RNA (mRNA), transfer RNA (tRNA), and ribosomal RNA (rRNA).

TABLE 1.1

Amino Acids, Their Three-Letter and One-Letter Abbreviations, and Codons

Amino acid	Three-letter	One-letter	Codons
Alanine	Ala	A	GCA GCC GCG GCU
Arginine	Arg	R	AGA AGG CGA CGC CGG CGU
Asparagine	Asn	N	AAC AAU
Aspartic acid	Asp	D	GAC GAU
Cysteine	Cys	C	UGC UGU
Glutamic acid	Glu	E	GAA GAG
Glutamine	Gln	Q	CAA CAG
Glycine	Gly	G	GGA GGC GGG GGU
Histidine	His	H	CAC CAU
Isoleucine	Ile	I	AUA AUC AUU
Leucine	Leu	L	UUA UUG CUA CUC CUG CUU
Lysine	Lys	K	AAA AAG
Methionine	Met	M	AUG (START CODON)
Phenylalanine	Phe	F	UUC UUU
Proline	Pro	P	CCA CCC CCG CCU
Serine	Ser	S	AGC AGU UCA UCC UCG UCU
Threonine	Thr	T	ACA ACC ACG ACU
Tryptophan	Trp	W	UGG
Tyrosine	Tyr	Y	UAC UAU
Valine	Val	V	GUA GUC GUG GUU
Stop codons			UAA UAG UGA

mRNA is transcribed from DNA and serves as the reading template from which proteins are created in a process called **translation**. tRNA carries amino acids to ribosomal complexes where protein translation is occurring. rRNA forms an organelle called a ribosome, which attaches to the mRNA template to decode the genetic information found there into amino acids, which form polypeptide chains.

DNA itself cannot be translated into protein; its instructions must be transcribed into mRNA, which serves as the protein translational template. The various codons within the DNA strand control how information is transcribed and which information is incorporated into mRNA. **Start codons** signal the beginning of a transcription sequence; **exons** are genomic sequences that should be incorporated into messenger RNA; **introns** are non-coding genomic sequence interspersed throughout the genome; and **stop codons** signal the end of the transcription process.

This unidirectional flow of information from DNA to RNA to protein, with DNA-to-DNA and RNA-to-RNA replication possible, is the fundamental tenet underlying modern biology and genetics and is called the **Central Dogma of Molecular Biology**.[5] Encyclopedic DNA, with all the protein-making instructions needed for replication, is transcribed to intermediate mRNA, which is the production template from which coded information is translated. RNA is then translated to protein. RNA-to-RNA replication is how viruses make copies of themselves (see Figure 1.2).

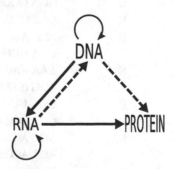

FIGURE 1.2
The Central Dogma of Molecular Biology showing RNA-to-RNA replication as the basis for how viruses make copies of themselves. (Source: Sage Ross/Wikimedia Commons/ Public Domain.)

The Virus as Industrial Insurrectionist

A virus is a type of *pathogen*—a disease-causing organism—that finds a way to enter host cells and then quietly takes over operating units within the cell to make copies of itself.

PATHOGEN VS. INFECTIOUS DISEASE

"Pathogen" and "disease" are not synonymous. A pathogen is, in the broadest sense, any organism that can produce disease. The term usually refers to microorganisms such as bacteria, fungi, or viruses (in contrast, small animals like worms or insects that cause disease are called parasites). The SARS-CoV-2 virus is a pathogen.

Infectious disease refers to the illness caused by a pathogen that can be transmitted from one person to another. When a new infectious disease arises, researchers begin searching for the pathogen that causes it. COVID-19 is the infectious disease caused by the SARS-CoV-2 virus.

If the host cell were an automobile factory making electric cars, the virus elegantly inserts itself as the new head of production, shuts down the assembly line making the electric autos, and redirects the plant to make gas-powered vehicles—all without alerting upper management. The virus acts like a stealth industrial insurrectionist, slipping unnoticed into a system and using it for its own agenda.

The "vehicle" of the manufactured virus has only two main parts: the "engine" is the *genome (DNA or RNA)*, and the "chassis" is the *membranous envelope* (see Figure 1.3).

The "factory" of the host cell has internal structures known as *organelles*, which are used for replication. The *ribosomes, endoplasmic reticulum, and Golgi body (apparatus, complex)* are especially important in the process[6] (see Figure 1.4). These three organelles are the primary components of the factory that the virus uses to replicate itself.

The virus carries instructions for making copies of itself within its RNA or DNA genome. Since viruses are simple machines, they only need to trick the host cell into replicating their two structural parts: the genome itself, with all the encoded information to make new viruses, and membranous envelopes in which to carry the genome.

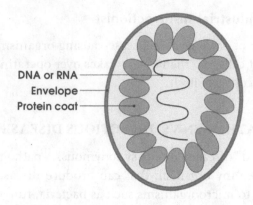

FIGURE 1.3

The DNA or RNA genome and envelope are two main parts of a virus. (Source: dom-domegg/Wikimedia Commons/licensed under CC BY 4.0.)

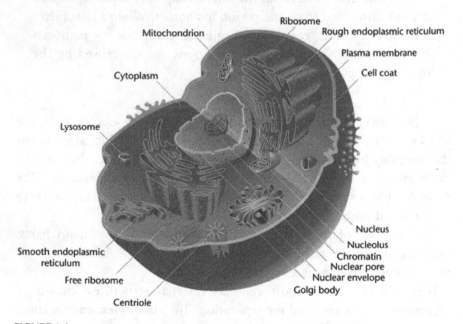

FIGURE 1.4

Internal structures of a cell showing ribosomes, endoplasmic reticulum, Golgi body, and other intracellular organelles that will be taken over and used for viral replication. (Source: Mediran/Wikimedia Commons/licensed under CC BY-SA 3.0.)

Viral replication is a factory production. Multiple copies of the viral genome and multiple copies of proteins forming the membranous envelope must be manufactured by the host cell, and then each of these component parts is assembled together. The manufactured end products made from instructions contained within the viral genome are proteins.

RNA or DNA is the genetic basis for propagating all living systems. Propagation requires the making of protein and assembling this protein into structures which comprise tissues and organs. Protein can only be translated from mRNA templates, and nuclear DNA (or RNA in other forms) must transition to mRNA for protein translation to commence.

Virus Classification Systems

While the cellular genetic material for higher-order cells is always double-stranded DNA, viral genomes have a variety of different genetic material configurations, including the type of genetic material comprising the genome (RNA or DNA), the conformation of this genetic material (single or double-stranded), and the orientation of the coding strand if the genome is single-stranded RNA (positive sense or negative sense).

In 1971 the biologist David Baltimore recommended using these genomic characteristics to categorize viruses. His categorization scheme came to be known as the **Baltimore system of virus classification** and assigns all viruses to one of seven categories based on their genetic material, conformation, and strand orientation:

- Double-stranded DNA (Group 1)
- Single-stranded DNA (Group 2)
- Double-stranded RNA (Group 3)
- Positive sense single-stranded RNA (Group 4)
- Negative sense single-stranded RNA (Group 5)
- Positive sense single-stranded RNA with DNA intermediates (retroviruses) (Group 6)
- Double-stranded DNA retroviruses (Group 7)

Another way to classify viruses is according to traditional biologic taxonomy, which specifically identifies and groups living organisms according to their shared morphology in a hierarchical manner: domain, kingdom, phylum, class, order, family, genus, and species. The **International Committee on Taxonomy of Viruses (ICTV)** is responsible for assigning names to new viruses (for instance, in March 2020 the ICTV assigned the name "severe acute respiratory syndrome coronavirus 2 or SARS-CoV-2" to what had been known as 2019 novel coronavirus[7]) and classifying viruses according to hierarchical taxonomy. ICTV classifies viruses into seven orders:[8]

- *Herpesvirales* are large eukaryotic double-stranded DNA viruses.
- *Caudovirales* are tailed ("caudal") double-stranded DNA viruses infecting bacteria.
- *Ligamenvirales* are linear double-stranded viruses infecting archaea (methane-producing microorganisms living in extreme environments on Earth, a form of life discovered in 1977).
- *Mononegavirales* are nonsegmented negative sense single-stranded RNA viruses infecting plants and animals.
- *Nidovirales* are positive sense single-stranded RNA viruses and are the order that includes coronaviruses like SARS-CoV-2.
- *Picornavirales* are small positive sense single-stranded RNA viruses which infect plants, insects, and animals.
- *Tymovirales* are monopartite positive sense single-stranded RNA viruses infecting plants.

Whatever form the viral genome assumes, once the virus attaches to a host cell and injects its genomic content into host cell cytoplasm, it must be transformed into mRNA to be translated by host cell ribosomes into protein. All mRNA is positive sense single-stranded RNA, and all viral genomes must become positive sense single-stranded RNA to begin protein translation. Viral genomes that are not conformed and oriented in this way (such as double-stranded RNA, single and double-stranded DNA, and negative sense genomes) require intermediate processing to develop messenger RNA templates, which can then be translated into the viral proteins needed to make new viruses. In contrast, the genomes of viruses that are already configured and oriented as positive sense single-stranded RNA—such as coronaviruses and other viruses within the *Nidovirales* order—can be translated immediately upon host cell entry. In terms of the industrial insurrection metaphor described above: the virus can only sneak into the factory and take over the manufacturing process if it is positive sense single-stranded RNA.

The SARS-CoV-2 Genome

The most important component of any virus is its genome, which contains all the information needed for the virus to make the proteins needed to replicate itself.

SARS-CoV-2 has 29 total proteins: 16 nonstructural proteins, four structural proteins, and nine accessory proteins. To make new virus,

SARS-CoV-2 will need to translate the coded information in its RNA, make an entire copy of the genome to be included in newly forming virus, and create mRNA templates to use for producing its 29 total proteins.

SARS-CoV-2 Is a Single-Stranded Positive Sense RNA Virus

SARS-CoV-2 is a **single-stranded positive sense RNA virus**. "Single-stranded" means the RNA genome is one continuous backbone of coded information (as opposed to a double-stranded structure of two continuous parallel backbones). "Positive sense" describes the orientation of this single strand of RNA from a starting point of protein translation called the 5-prime end (5'end) to an ending point called the 3-prime end (3'end).

The single-stranded positive sense RNA genome has its genetic information ordered in such a way as to begin translating proteins needed in choreographed steps to take over host cell machinery and begin the viral replication process[9] (see Figure 1.5).

To return to the metaphor of industrial insurrection: a single-stranded positive sense RNA virus arrives at the automobile factory with everything needed to immediately switch the assembly line from making electric cars (copies of the host cell) to producing gas-powered vehicles (copies of the virus), without causing a single hiccup in the assembly line.

Protein Translation Starts with ORF1a Making NSP1–NSP11

The first step in viral takeover of host cell machinery is having host ribosomes recognize the SARS-CoV-2 positive sense messenger RNA template and begin translating viral genetic instructions into viral proteins.

Host cell ribosomes latch onto the untranslated region (UTR) of the SARS-CoV-2 RNA genomic strand at the 5' starting point and move without stopping along an **open reading frame (ORF)** with coded instructions for making proteins. An ORF allows unfettered movement along genomic templates.

For the SARS-CoV-2 genome, the first opening reading frame, ORF1a, is translated into an elongated protein chain called polyprotein pp1a (pp1a) containing 11 nonstructural proteins (NSPs): NSP1 through NSP11. NSPs are essential for making component parts of new viruses, but as

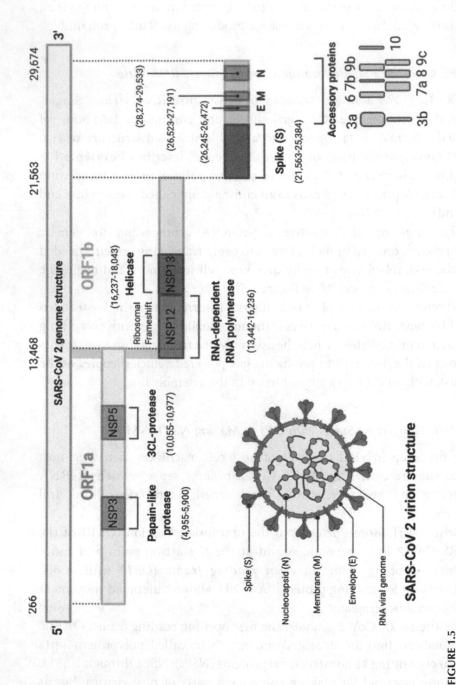

FIGURE 1.5

SARS-CoV-2 genome showing open reading frames, nonstructural proteins, structural proteins, and accessory proteins, accompanied by SARS-CoV-2 virion structure. (Source: Alanagreh et al., Pathogens, 2020.)

nonstructural proteins they are not assembled as nascent proteins within newly made viruses.

The most notable NSPs in this first polyprotein chain are NSP1, NSP3, and NSP5. NSP1 shuts down the host cell's usual protein translation to enable viral protein translation, which is likely why NSP1 is the very first nonstructural protein to be translated. In the industrial insurrection metaphor, this can be thought of as the virus switching the parts coming down the assembly line from electric vehicle parts to gas-powered car parts. The organelles working in the factory simply continue their usual production process—but now they're working on the virus' material.

Viral proteins needed for replication cannot perform their functions until being cut away from polyproteins. After NSP1 enables viral protein translation to begin, NSP3 (the papain-like protease) and NSP5 (the *3CL-protease or main protease M^{pro} (NSP5)* come into play. These are both proteases, which are needed to cleave the elongated chains of inactivated polyproteins pp1a and pp1ab into active proteins necessary for continuing viral replication.

Protein Translation Continues with ORF1b Making NSP12–NSP16

Ribosomal translation continues from ORF1a to another opening reading frame: ORF1ab. This stage of protein translation produces a second elongated protein chain, polyprotein pp1ab (pp1ab), which contains five more NSPs: NSP12 through NSP16.

The most important nonstructural protein at the beginning of pp1ab is NSP12, known as *RNA-dependent RNA polymerase (RdRp; NSP12).* All RNA viruses produce RdRp, which allows the kind of RNA-to-RNA transfer of information described by the Central Dogma.

RdRp creates two different types of RNA through its transcription activities, which allow RNA-to-RNA information transfer:

- RdRP transcribes a copy of the entire single-stranded positive sense RNA genome, which will eventually be encapsulated within newly forming viruses called "copy genomic RNA."
- RdRp also transcribes fragmented copies of messenger RNA (called "subgenomic RNA"), which are used as translation templates by host cell ribosomes to make four structural proteins comprising SARS-CoV-2: spike (S), nucleocapsid (N), membrane (M), and envelope (E).

Now the metaphorical auto plant is making both the virus' engine (the RNA genome) and the chassis (the membranous envelope). Rolling off the end of the assembly line are new viruses, ready to drive off into the host organism, slip into other factories, and begin the process anew.

Immune System Evasion and Accessory Proteins

Any insurrectionist must evade detection by authorities that may halt its activities. As a virus goes about its takeover of the metaphorical auto plant, why don't the "overseers"—the immune system that identifies and attacks foreign invaders—put an immediate stop to it?

First, the incredibly simple structure of a virus makes it more difficult to identify as an invader when it enters the body—unlike bacteria or fungi, whose cellular structure is so different from human cells that the immune system easily identifies and attacks them. A virus slips into the body like James Bond in a membranous envelope tuxedo.

Second, the replication process happens inside a host organism's cells, where it is difficult to detect. Imagine the immune system as a board of overseers sitting high above the factory floor. From that vantage point, the factory seems to be functioning as it should, and the components of the gas-powered cars (the elements of the virus) look very similar to the electric vehicle components (the host cell's own proteins) that the factory is supposed to be working with.

Finally, accessory proteins within the virus' genome itself help prevent detection by the immune system. In the case of SARS-CoV-2, the RNA genome contains nine accessory proteins (ORF3a, ORF3b, ORF6, ORF7a, ORF7b, ORF8, ORF9a, ORF9b, and ORF10), which have no direct role in viral replication and are thought to aid immune system evasion.

The immune system does eventually uncover the plot thanks to major histocompatibility complexes (MHCs)—like an elite anti-espionage team within a cell—which we will discuss in Chapter 7: Vaccines.

SARS-CoV-2 Life Cycle

SARS-CoV-2 enters cells through the **angiotensin-converting enzyme 2 (ACE2)** receptor found throughout the body, most prominently in nasopharyngeal, respiratory, and gastrointestinal tissues. This **tropism**, or preferential tendency for SARS-CoV-2 to use ACE2 as an entry point into certain cells, is a likely explanation for characteristic COVID-19 signs and symptoms, including loss of smell and taste, pneumonia, and diarrhea.

The ACE2 receptor is an important modulating component of the ***renin-angiotensin-aldosterone system (RAAS)***, a critical physiologic mechanism maintaining fluid balance and blood pressure control. The disruption caused by SARS-CoV-2 binding to the ACE2 receptor has a cascading effect, leading to immune system dysregulation, thromboembolism, and acute respiratory distress syndrome, which may be fatal.

SARS-CoV-2 Cell Entry, Cytoplasmic Replication, and Exit

Cellular proteases act on the spike glycoprotein, which binds with ACE2 through an S1 subunit and fuses and enters the cell through an S2 subunit. The S1 subunit is called the receptor binding domain and is the protein component with the largest number of mutations, allowing SARS-CoV-2 to evade antibodies targeted against it (see Figure 1.6).

After cell entry through ACE2, the virus uncoats itself, releasing its single-stranded positive RNA genome, and ***cytoplasmic reproduction*** begins—the metaphorical takeover of the factory. The four structural proteins are assembled along with copies of genomic RNA within the intermediate space of the endoplasmic reticulum—Golgi interface (ERGIC) and enveloped into budding vesicles. Finally, the newly minted copy of the virus, known as the ***virion***, exits the host cell through exocytosis.

Interrupting the Virus Life Cycle

At each stage of the virus life cycle, as well as within the choreographed steps of the cytoplasmic reproduction stage, there are opportunities for detection and attack by both the body's internal defenses and therapeutic agents. SARS-CoV-2 has, thus far, been most effectively targeted by medications during the cytoplasmic reproduction stage, which are discussed in more detail in Chapter 5: Therapeutics. Examples include the following.

- M^{pro}: This protease is the target for a COVID-19 medication called nirmaltrelvir, which stops M^{pro}'s ability to cut polyproteins into active fragments.

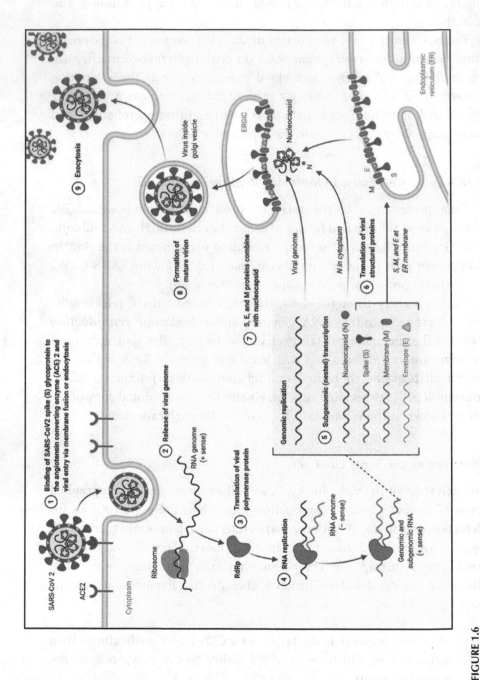

FIGURE 1.6

SARS-CoV-2 life cycle showing cell entry using ACE2 receptor, cytoplasmic replication, viral assembly, and virion exit by exocytosis. (Source: Alanagreh et al., Pathogens, 2020.)

- RdRp: The role of RdRp is so essential to viral replication that any interruption in its actions will prevent creation of new viruses. Two COVID-19 medications, remdesivir and molnupiravir, disguise themselves as nucleotides that would normally be incorporated by RdRp in its transcription activities creating new RNA forms, but due to their inauthenticity both medications halt the actions of RdRp and prevent new virus formation.

Virus Mutability

Viruses satisfy the second requirement for NASA's definition of life, as they are capable of Darwinian evolution. Existing under constant attack from host cell immunologic defense mechanisms (antibodies, macrophages, lymphocytes, and other immune system elements) and human-manufactured medical countermeasures (vaccines, monoclonal antibodies, and therapeutic agents) requires all viruses to alter their virion structures to survive. Alteration of the virion structure occurs in two ways: within the viral genome itself through mutations or through recombination, where two strains of virus infecting the same host cell mix their genetic material to create new viral progeny, which may become a new lineage of virus.

How Mutations Occur

As translation begins, host cell ribosomes may incorrectly read genetic information contained on the mRNA template, substituting or deleting amino acids based on this misread information. To counteract this, some viruses have proofreading ability to excise aberrant sections through exonuclease enzymes within their genomes, and this beneficial proofreading ability preserves fidelity of the genome and its genetic information.

But if a virus does not possess proofreading ability, these misreads become mutations, which permanently alter the virus's genetic code. If the mutations occur in parts of the viral genome that dismantle essential replication functions, virion creation cannot continue, and the misread mutations end there. But if the mutations create protein configurations that make the virus more transmissible or allow it to evade immune detection, these "good" mutations confer an evolutionary advantage to the virus and are retained and propagated within the genome. In other words, genotypic mutations that produce phenotypic advantages are preserved in the viral genome and passed on to virion progeny, and the culminating success of all these changes can produce a new strain of the original virus which is better equipped for survival.

Imagine the industrial insurrectionist again. It has successfully infiltrated the automobile factory, reprogrammed the machinery to make gas-powered vehicle components, and the production line is humming along. Then the machine that stamps metal into mufflers has a little glitch and spits out a misshapen muffler—a mutation has occurred. The mutated part turns out to fit right in with the rest of the car's components—in fact, it makes the engine function a little better—and a new variant rolls off the end of the assembly line.

Mutation Terminology

Various terms are used to describe how viruses modify themselves in response to environmental pressures, and it is important to clarify the proper use of these terms.

- **Mutation**: a *mutation* is the actual change in the amino acid sequence within the genome. Amino acids may be substituted into the growing polypeptide chain or entirely deleted out based on ribosomal misreading of the mRNA template. A shorthand way of describing the amino acid changes of a mutation is by using a one-letter abbreviation for the amino acid appended to the position number in the genome sequence where the change occurred. These one-letter abbreviations are shown in Table 1.1. For instance, one of the earliest and most successful mutations of the ancestral SARS-CoV-2 genome was D614G, in which the amino acid aspartic acid (D) is replaced by glycine (G) at the 614th position of the SARS-CoV-2 spike protein's 1,273 genome sequence. The 1,273-spike protein genome sequence has a spike S1 attachment domain from position 1 to 686, including the receptor binding domain from position 306 to 534, and a spike S2 fusion domain from position 687 to 1,273. The D614G mutation created a new spike protein configuration, giving it stronger affinity and firmer adherence to the ACE2 receptor of host cells.

- **Variant**: a *variant* is an entirely new genome sequence resulting from one or more mutations. Emerging SARS-CoV-2 variants amass a set of mutations, which collectively provide greater transmissibility, immune evasion, and survivability. The Centers for Disease Control (CDC) and World Health Organization (WHO) prioritize these variants according to a hierarchy of concern—from Variants Being Monitored (VBMs) to Variants of Concern (VOCs)—based on their ability to evade medical countermeasures.
- **Lineage**: a *lineage* is a distinct branch on the virus's phylogenetic tree determined by the new variant's genetic sequencing. SARS-CoV-2 lineages are classified in two ways: one is the *Phylogenetic Assignment of Named Global Outbreak (PANGO) lineage* system, which assigns an alphabetical prefix and up to three numerals separated by periods to indicate sub-lineages, and the other are *WHO Greek letter labels* calling a particular lineage by a Greek letter. For instance, the Delta (WHO Greek letter label) variant is comprised of B.1.617.2 and AY lineages (PANGO lineage).

SARS-CoV-2 Variants

Global genomic surveillance programs track and follow emerging SARS-CoV-2 variants. These may crop up at any time, in any country or region, and are usually indicated by PCR testing where an *S-gene target failure (SGTF)* occurs during the laboratory assay. PCR tests target several nucleic acid components of the SARS-CoV-2 virus, and when an expected reading from spike protein (S-gene) is not present, this may indicate a mutation within the spike protein itself.

The Alpha variant, also known as B.1.1.7, emerged in September 2020 and became the dominant circulating form of SARS-CoV-2 in England before it traveled to other countries and dominated new infections in these regions. Since the emergence of Alpha, there have been four other major Variants of Concern: Beta (B.1.351) originating in South Africa in October 2020, Gamma (P.1) arising first in Brazil and Japan in January 2021, Delta (B.1.617.2) identified in July 2021, and Omicron (B.1.1.529) emerging in November 2021. These variants share several of the same mutations (e.g., nearly all contain D614G), as well as having their own distinct mutations. Omicron is the most virulent and transmissible SARS-CoV-2 variant to date, with over 50 mutations, including at least 30 amino acid substitutions, three deletions, and one amino acid insertion—all conferring increased transmissibility to Omicron.

More Variants: Not If, but When

SARS-CoV-2 variants will continue to emerge, and infect vaccinated and unvaccinated people—making unvaccinated and previously unexposed persons sicker with their infections—and transition to new lineages as long as susceptible hosts are available to be infected. The demonstrated ability of SARS-CoV-2 to mutate and produce more virulent variants all but guarantees that this virus will continue to evolve.

The CDC's advice to "stay up to date with your vaccines" means many vaccinated people had already received four shots against SARS-CoV-2 by spring 2022: two primary series shots and, depending on age and host factors, two booster doses. Fortunately, mRNA technology for vaccine development has been able to keep pace with emerging variants as global genomic surveillance continues. SARS-CoV-2 will very likely succumb to the fate of influenza and other respiratory viruses by becoming an endemic pathogen, mutating and making its worldwide sweep each season.

Looking Back, Looking Ahead

On a biological level, a fundamental action of all living organisms is the perpetuation of their genome. While viruses are not self-sustaining, they usurp the self-sustaining mechanisms of other organisms to sustain themselves. In pursuing this goal, they can cause disease—the destruction of host cells and other effects that give rise to illness and death.

The function and life cycle of viruses is well understood, and identifying the specific characteristics of the SARS-CoV-2 virus has enabled the creation of therapeutic agents and vaccines to combat COVID-19. These principles provide the necessary foundation for understanding how this novel virus arose, spread, and caused disease—as well as how scientists developed therapeutics and vaccines to combat it.

Though the SARS-CoV-2 virus does not meet NASA's condition of "self-sustaining" to be considered a type of life, it has nevertheless sustained itself. Indeed, it has thrived—replicating, spreading, and mutating at a rapid pace. The persistence of this pathogen is attributable to both its specific ability to replicate itself within host cells and the larger environment in which it exists. An interesting feature of the global COVID-19 pandemic is the vast array of cultural ideas—from conspiracy theories to social conventions to beliefs about individual and collective rights and responsibilities—that arose and were sustained along with the virus itself. In both cases, the mechanism for sustaining this "life" was the same: the human host.

FUTURE OUTINGS

To understand how the lessons of COVID-19 can be applied to future EIDs, ongoing excursions into the field must explore:

- The precise functions of SARS-CoV-2 accessory proteins, especially as they relate to immune system evasion
- Additional opportunities for therapeutic agents to disrupt the virus life cycle, especially in light of new SARS-CoV-2 variants

Student Research Questions

1. How do single-stranded positive sense RNA viruses differ from other virus structures in terms of ability to evade the immune system, replicate, and spread?
2. How was the SARS-CoV-2 genome identified and sequenced? What organizations and individuals contributed to this work?
3. What other single-stranded positive sense RNA viruses exist? How well are their genetic structures understood? In what ways are they similar to and different from SARS-CoV-2?
4. If an organism is not "self-sustaining," what opportunities and challenges does that provide for medical science to combat it?
5. Are there other organisms or entities that sustain themselves through external means, like viruses? How do viruses compare in this way to abstract ideas and cultural beliefs? What ideas and beliefs were "sustained" in the culture during the COVID-19 pandemic, and how did they replicate themselves?

NOTES

1. On July 20, 1969, seven years before the Viking missions, Apollo 11 landed on the moon. Though a milestone achievement in space exploration, a moon is not a planet (moons orbit around planets; planets orbit around stars like the sun), so the Viking missions to Mars were humankind's first interplanetary expedition.
2. "What is astrobiology?," G. Scott Hubbard (October 1, 2008) edited by Brian Dunbar (August 7, 2017), https://www.nasa.gov/feature/what-is-astrobiology, accessed May 2, 2022.
3. "Defining life," Steven A. Benner, Astrobiology, volume 10, number 10, 2010, pages 1021–1030, doi:10.1089/ast.2010.0524.
4. Hua X, Bromham L, "Darwinism for the genomic age: connecting mutation to diversification." Frontiers in Genetics. February 2017. 8:12. doi:10.3389/fgene.2017.00012.
5. Cobb M (2017) 60 years ago, Francis Crick changed the logic of biology. PLoS Biol 15(9): e2003243. https://doi.org/10.1371/journal.pbio.2003243, accessed May 5, 2022.
6. "Golgi body," "Golgi apparatus," and "Golgi complex" are all synonymous terms for this intracellular organelle with an important role in protein assembly.
7. Gorbalenya AE et al. March 2020. "The species *Severe acute respiratory syndrome-related coronavirus:* classifying 2019-nCoV and naming it SARS-CoV-2." Nature Microbiology, volume 5, pages 536–544. https://doi.org/10.1038/s41564-020-0695-z.
8. Gita Mahmoudabadi, Rob Phillips (2018). Research: A comprehensive and quantitative exploration of thousands of viral genomes. eLife7:e31955. https://doi.org/10.7554/eLife.31955.
9. Alanagreh, Lo'ai, Foad Alzoughool, and Manar Atoum. 2020. "The Human Coronavirus Disease COVID-19: Its Origin, Characteristics, and Insights into Potential Drugs and Its Mechanisms" Pathogens 9, no. 5: 331. https://doi.org/10.3390/pathogens9050331.

2

The Origins of SARS-CoV-2

The Field: In order to respond to a new pathogen, scientists must understand its origins. Less than two months after a new, unknown pathogen began infecting people in a large city in China, scientists identified the virus as SARS-CoV-2, a new type of coronavirus—but more than two years later, debate continues about how that virus came to exist. Understanding the ideas behind the two prevailing theories—and the forces that have framed the debate—is pivotal in determining the origins of SARS-CoV-2, as well as how the investigation into future disease origins can be effectively conducted.

Field Sightings: causative agent, Wuhan Institute of Virology's (WIV), emerging infectious diseases (EIDs), zoonotic pathogens, zoonotic spillover theory, intermediate host, animal reservoir, maintenance host, laboratory leak theory.

FIELD EXPEDITION: THE BAT WOMAN

Virology is an obscure branch of microbiology, and virologists are not well-known public figures. But one virologist, Dr. Shi Zhengli, found herself at the center of one of COVID-19's biggest controversies: whether its *causative agent*, the SARS-CoV-2 virus, came from an animal source or was in fact released from the *Wuhan Institute of Virology's (WIV)* National Biosafety Level 4 Laboratory—the very laboratory where she leads a team of researchers investigating *emerging infectious diseases (EIDs)*.

Shi has been affectionately called "the bat woman" by her colleagues because of her pioneering work with bat species. After receiving her PhD

DOI: 10.1201/9781003310525-3

from Université Montpellier 2 in 2000, she began investigating viruses in shrimp and crabs, but this work was soon diverted by the 2002 SARS outbreak, which eventually sickened 8,098 people worldwide, causing 774 deaths.[1]

China and Hong Kong were the two global regions most affected by SARS. Of the 8,098 people who contracted SARS, 5,327 were in China and 1,755 in Hong Kong. Among the 774 SARS-related deaths, 349 were in China and 299 in Hong Kong.[2] SARS was principally a Chinese infectious disease outbreak, and by 2004 the Chinese Academy of Sciences instructed the WIV, where Shi had been working since 1990, to lead the investigation into the origins of SARS. Shi and other WIV scientists were the first to prove that bats were a natural reservoir for SARS-like coronaviruses in a landmark 2005 study.[3] WIV was also the first to sequence the genome of SARS-CoV-2, publishing these results on February 3, 2020,[4] which enabled Moderna to start human clinical trials on its mRNA vaccine by March 16, 2020.

Shi continued her work collecting bat saliva, blood, urine, and fecal samples to find links between bats and coronaviruses, eventually amassing some 20,000 bat specimen samples. Because of her extensive field and lab activities—and the important scientific discoveries they enabled—Shi's moniker as "the bat woman" is well-earned. That bats are seen as a symbol of good luck and blessing in Chinese culture further underscores the apt nature of Shi's playful nickname.

But anti-Chinese sentiment increased dramatically on March 16, 2020, when President Trump began referring to SARS-CoV-2 not by its taxonomic designation but instead calling it "the Chinese virus."[5] One month later, Washington Post columnist Josh Rogin published an opinion piece about alleged safety concerns at WIV, giving credence to those expounding a theory that SARS-CoV-2 entered the world by leaking from the WIV.

LESSON LEARNED: THE BAT WOMAN

Shi became the visible face of anti-Chinese sentiment as suspicions were raised about the work of her WIV laboratory. Despite counterarguments made by Shi of no genetic relationship between SARS-CoV-2 and bat coronaviruses studied in her lab, along with evidence that WIV lab personnel showed no antibody signs of coronavirus exposure, a "lab leak" was promoted as a possible cause of SARS-CoV-2.

The fond characterization of "the bat woman" became twisted into a sinister depiction of Shi and her research team at the WIV. That her pioneering research was re-framed as a source of suspicion displays how the origins of the SARS-CoV-2 virus were politicized and Chinese scientists were vilified as the world raced to understand the true source of this new virus.

CONCEPTS COVERED IN THIS CHAPTER

- Emerging infectious diseases (EIDs)
- Zoonotic spillover theory
- Bats as animal reservoirs
- Laboratory leak theory
- Debate regarding laboratory leak and zoonotic spillover theories

SARS-CoV-2: An Emerging Infectious Disease

The global COVID-19 pandemic began as a set of pneumonia cases caused by an unknown agent among residents of Wuhan, the capital city of central China's Hubei Province—a huge city with a population of approximately 11 million people, three million more than New York City.

The first cases of pneumonia were noted in December 2019, but earlier cases likely occurred and went undetected.[6] By January 3, 2020, 44 patients who had contracted pneumonia of unknown cause were reported to the World Health Organization (WHO). When these first 44 cases were reported to the WHO, 33 patients were in stable condition and 11 had already become seriously ill—one of whom died within the following month.[7] The outbreak seemed to be associated with an animal market in Wuhan, which was promptly closed. By January 12, Chinese authorities had identified, isolated, and sequenced a new type of coronavirus presumed to be the causative agent of the deadly pneumonia, ruling out other known respiratory pathogens such as influenza, avian influenza, adenovirus, and other known coronaviruses.[8]

The unfolding story of COVID-19 is typical of an EID: symptoms arising in a small or localized population for which a clear pathogen cannot be immediately determined, followed by spread of the disease and research into and eventual identification of the pathogen behind it. Because SARS-CoV-2 emerged in a large, densely populated city—one that serves as a travel hub among many other large cities—the spread of the EID was both broad and rapid.

Other EIDs afflicting human populations in recent years include acquired immunodeficiency syndrome (AIDS), severe acute and Middle East respiratory syndromes (SARS and MERS, respectively), the hemorrhagic fevers Marburg and Ebola, and Rift Valley fever.[9]

What characterizes EIDs as "emerging" is how they arise, which can occur in four ways:

- Through newly discovered pathogens
- Through a known pathogen appearing in a new geographic area
- Through a known pathogen reappearing in an area where it caused previous infection
- Through a known pathogen becoming antibiotic resistant[10]

Early in its course, the pathogen responsible for COVID-19 was identified as a "novel" or new type of coronavirus. Four coronaviruses cause common colds or mild respiratory illnesses: HCoV-229E, HCoV-OC43, HCoV-NL63, and HCoV-HKU1. This new coronavirus—first called "2019 novel coronavirus" (2019—nCoV) and later "severe acute respiratory syndrome coronavirus 2" (SARS-CoV-2)—clearly had more serious implications than the coronaviruses listed above. At the beginning of the twenty-first century, two coronaviruses had caused serious human illness; they were responsible for the 2002–2003 SARS and 2012 MERS epidemics.[11] In 2020, SARS-CoV-2 became the third coronavirus to be responsible for a global health crisis.

Scientists identified the new pathogen less than two months after symptoms of the disease were first observed—but where did SARS-CoV-2 come from? This question has been a source of controversy from the start of the pandemic to the present, with dissenting views among scientists, politicians, and government leaders framing the debate about its origins. Two camps quickly emerged: those pointing to an animal source—as had been the case for the majority of EIDs in the past—and those claiming that the pathogen was created by human beings through laboratory experiments.

Zoonotic Spillover Theory

Approximately 60% of EIDs are estimated to be caused by *zoonotic pathogens*. These microorganisms are often viruses from vertebrate animals, and 75% of them spill over, or "jump species," to humans.[12,13,14] This zoonotic spillover is no easy task for a pathogen and *zoonotic spillover theory* holds that several factors must align before spillover from vertebrate animals to humans can occur, including:

- Pathogen prevalence, release, and survival
- Pathogen amplification and spread, sometimes involving an *intermediate host* or other animal source
- Overcoming human host immunologic barriers after exposure[15]

Bats have been implicated as a common *animal reservoir* responsible for transmitting a number of emerging infectious diseases. Rabies is the longest studied virus known to have arisen from bats,[16] and more recently Ebola, Marburg, Nipah, Hendra, and the coronaviruses linked to SARS and MERS have all been associated with bats as an originating source. The zoonotic spillover theory holds that SARS-CoV-2 evolved from a bat coronavirus and was transmitted to humans through an animal intermediary, similar to the coronaviruses that caused the SARS and MERS epidemics (see Figure 2.1).

Intermediate animal hosts support and amplify zoonotic pathogens, which spill over to humans. For example, the palm civet is considered the intermediate animal host for the SARS coronavirus and the dromedary camel for MERS coronavirus. Pangolins were at one point believed to have been the intermediate animal host for SARS-CoV-2, but an analysis of species sold in Wuhan did not find any pangolins among some 47,000 animals sold in its animal markets from May 2017 through November 2019 before the outbreak of COVID-19.[17]

Though an intermediate host has yet to be identified for COVID-19, bats are the most likely animal reservoir for any progenitor viruses giving rise to SARS-CoV-2.

Bats as Animal Reservoirs for Lethal Zoonotic Pathogens

Over 52 million years ago, bats evolved from a common ancestor into two suborders, *Yinpterochiroptera*, previously known as the "megabats," and *Yangochiroptera* or "microbats." The taxonomic classification order of bats

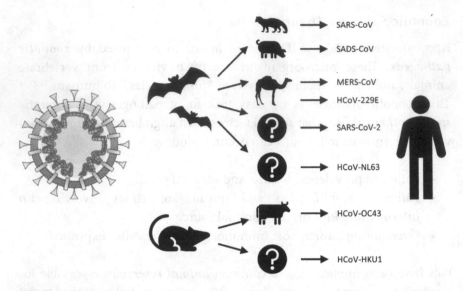

FIGURE 2.1
Animal origins of human coronaviruses. (Source: nagualdesign/Wikimedia Commons/licensed under CC BY 4.0.)

is *Chiroptera,* from the Greek "chiro," which means "hand," and "pteron," meaning "wing." The "handwing" of bats allows powered flight, and bats are the only mammalian species endowed with this ability. Indeed, the energy expenditure and metabolic requirements of mammalian powered flight are likely important catalysts for several cellular mechanisms and immunologic responses that have evolved within bats. These adaptive mechanisms may be contributing factors to bats' ability to coexist with a number of virus species known to cause emerging, and in some cases lethal, human infectious diseases.

With some 1,200 species, bats are the second-most abundant mammalian order, second only to *Rodentia*, the largest mammalian about comprising 1,500 species (including rats, mice, hamsters, guinea pigs, beavers, and squirrels). Bats and rodents are both important sources for infectious diseases, but researchers have focused intense interest on bat species as a leading source of zoonotic illness due to their:

- Extensive geographic range
- Large-colony roosting behavior
- Relatively long lifespan compared to other small mammals
- Coevolution with viral species over millions of years, which may be related to the metabolic requirements of mammalian powered flight

Bats can be found living in all terrestrial habitats except the Antarctic, and are generally found roosting together in colonies as large as ten million individuals. Bats give birth once a year to one or two offspring and may live as long as 30 years, compared to mice which may live up to four. These characteristics give bats ample opportunities to encounter and spread pathogens to intermediate animal hosts and humans.

Bats seem to have a particular ability to carry lethal viruses without incurring illness from those pathogens, thereby acting as maintenance hosts that can transmit diseases to other mammals. Lyssaviruses, which cause rabies, were first associated with bats as an animal reservoir. Since then, many zoonotic pathogens have been associated with bats as a *maintenance host*, including henipaviruses (Hendra and Nipah viruses), filoviruses (Ebola and Marburg viruses), and, most recently, the coronaviruses causing SARS, MERS, and, possibly, SARS-CoV-2.[18]

Although the precise mechanisms by which bats are able to "peacefully coexist" with lethal viruses is yet to be determined, a number of hypotheses have been offered to explain this phenomenon:

- Powered flight requires a great deal of metabolic energy, causing bats to increase mitochondrial efficiency within their cells.
- The accumulation of metabolic byproducts from oxidative damage may have stimulated the immune systems of bats to evolve in ways that keep viruses in check.
- Bats and the viruses residing within their colonies, especially RNA viruses with high mutation ability, have coevolved over millions of years, allowing for fine-tuning of mechanisms that allow for their coexistence.

Laboratory Leak Theory

As COVID-19 emerged and spread, speculation about its origins was rapid and rampant. In stark opposition to the zoonotic spillover theory, an opposing viewpoint emerged that placed the origin of the virus squarely in the human realm: the *laboratory leak theory*.

One month before the WHO declared COVID-19 to be a global pandemic,[19] United States Senator Tom Cotton of Arkansas was already proposing a possible origin for SARS-CoV-2:

> we don't know where it originated, and we have to get to the bottom of that … we also know that just a few miles away from that food market is China's

only biosafety level 4 super laboratory that researches human infectious diseases.[20]

Based on this correlation, Cotton promoted a laboratory mishap as the source for SARS-CoV-2 over the initial, and still prevailing, consensus that virus transmission from animals in Wuhan's live animal food markets spawned the newly lethal respiratory pathogen. Cotton's "fringe idea"[21] can be considered part of the "laboratory leak" theory.

The "laboratory leak" theory contends that scientists within the Wuhan Institute of Virology's (WIV) National Biosafety Level 4 Laboratory, who were possibly involved in gain-of-function research[22] on bat coronaviruses, may have either inadvertently or purposefully released modified coronavirus specimens into the surrounding community, causing the first cases of COVID-19. Support for the idea of an artificial origin for SARS-CoV-2[23] was pointedly stated in a May 2021 letter to *Science* magazine co-signed by 18 scientists:

> As scientists with relevant expertise, we agree with the WHO director-general, the United States and 13 other countries, and the European Union that greater clarity about the origins of this pandemic is necessary and feasible to achieve. We must take hypotheses about both natural and laboratory spillovers seriously until we have sufficient data. A proper investigation should be transparent, objective, data-driven, inclusive of broad expertise, subject to independent oversight, and responsibly managed to minimize the impact of conflicts of interest.[24]

However, an opposing group of 22 virologists and scientists made the following arguments against the "lab leak" theory in a September 2021 article in the journal *Cell*.[25]

- **No likely SARS-CoV-2 progenitor viruses studied at WIV**: the bat coronavirus being studied at WIV was RaTG13. There is an approximate 4% genetic difference between RaTG13 and human coronavirus SARS-CoV-2, which is similar to the genetic difference between humans and chimpanzees[26]—a significant gap. Genetic recombination among three other bat viruses (RmYN02, RpYN06, and PrC31) are more closely related to human SARS-CoV-2 than RaTG13 and would therefore be more likely progenitors. However, WIV did not study or collect samples of these other bat viruses.

- **Viral inactivation**: WIV and other laboratories routinely use the technique of viral genomic sequencing without cell culture, in which viruses are inactivated during RNA extraction, making WIV incapable of releasing a live virus with infection potential.
- **Negative serologic studies**: WIV scientists and staff were reported to be seronegative for SARS-CoV-2 when tested in March 2020; if these workers had been handling SARS-CoV-2 as part of experimentation, serology studies would likely be positive.
- **No evidence of artificial genetic markers**: as the authors stated, "SARS-CoV-2 carries no evidence of genetic markers one might expect from laboratory experiments."
- **No clear means for a virus to escape**: a laboratory escape scenario in which infected mice made their way out of the facility is invalidated by the fact that early forms of SARS-CoV-2 were not able to infect mice.
- **Mutability of SARS-CoV-2**: claims that the virus was engineered to be easily transmissible to humans are countered by the evidence that, as the pandemic progresses, SARS-CoV-2 continues to demonstrate a remarkable ability to mutate, increasing its transmissibility and virulence to a degree such that SARS-CoV-2 would not need "'humanized' animal models to promote human transmission."

Official Reviews of Laboratory Leak and Zoonotic Spillover Theories

In January 2021, the WHO released its "WHO-convened Global Study of Origins of SARS-CoV-2: China Part, Joint WHO-China Study, 14 January–10 February 2021, Joint Report." This voluminous publication proposed that further inquiry into both lab leak and zoonotic spillover theories be conducted to determine the probability of one of four scenarios:

1. Direct zoonotic transmission
2. Intermediate host followed by zoonotic transmission
3. Cold/food chain transmission
4. Laboratory incident as a source of transmission

The report was not well-received, as claims of WHO bias and Chinese obfuscation were levied against it.

In late May 2021, the United States government entered the fray when President Biden issued a statement saying that "the U.S. Intelligence Community has coalesced around two likely scenarios." In this statement, the President directed the intelligence community to further investigate whether the virus "emerged from human contact with an infected animal or from a laboratory accident" and to "bring us closer to a definitive conclusion, and report back to me in 90 days."[27] As a result, on October 29, 2021, the declassified report "Office of the Director of National Intelligence, National Intelligence Council: Updated Assessment on COVID-19 Origins" was released with the following conclusions.

- **Areas of broad agreement**: SARS-CoV-2 was not developed as a biological weapon, was not genetically engineered, and China's officials were unaware of the virus before the pandemic emerged.
- **Two plausible hypotheses on initial human exposure**: natural transmission from animal to human or laboratory-associated incident were both plausible, but evidence was not strongly diagnostic of either hypothesis.
- **China's cooperation key to understanding origins**: Beijing's lack of cooperation with other countries and international groups has impeded investigation into the origins of SARS-CoV-2; China's participation in bridging numerous information gaps, particularly related to technical data, is crucial to confirm either hypothesis.

A natural animal-based source, especially one implicating bats as the animal reservoir for a progenitor virus, has been—and remains—the prevailing hypothesis for the origin of SARS-CoV-2 among the scientific community. However, given the partisan nature of the continuing dispute between "lab leak" and zoonotic spillover theories, the global community may not reach broad consensus any time soon.

Looking Back, Looking Ahead

When a new EID appears, it presents a scientific mystery that must be solved in order to develop effective tools for diagnosing, treating, and preventing further transmission of the disease. While there is no question regarding where SARS-CoV-2 originated—and the role that Wuhan's large population and global interconnectedness played in the spread of the virus—the source of the pathogen itself has not been definitively determined. The scientific consensus points to an animal source; similar to the other epidemic-causing coronaviruses, SARS-CoV-2 has likely been harbored by bats as reservoir with an unidentified animal intermediate host.

Yet the lab leak theory persists among some scientists and political figures, and important international organizations like the WHO continue to entertain both hypotheses. The location in which SARS-CoV-2 initially emerged is itself a complicating factor: China's internal policies and the attitudes of various countries and political figures toward China all factor into how the investigation into the virus' origins has unfolded.

Answering the question of the origin of SARS-CoV-2 is critically important. If SARS-CoV-2 originated as another example of an EID caused by an animal source, how should other pandemic-potential pathogens be better handled in the future? If SARS-CoV-2 originated from a foreign government mishap, either as an intentional act of bioterrorism or as an accidental event, how should the world respond? Finally, how can scientists effectively determine the origins of future EIDs in the face of political factors that influence investigation?

FUTURE OUTINGS

To understand how the lessons of COVID-19 can be applied to future EIDs, ongoing excursions into the field must explore:

- The specific species of animals that may have served as the animal reservoir and intermediate host of SARS-CoV-2
- The methodology behind the laboratory leak theory and the validity of its claims
- The effect of China's relationships with other countries and the global community on the development of and investigation into SARS-CoV-2 origin theories

Student Research Questions

1. What role have zoonotic pathogens played in past epidemics and pandemics throughout history? As human populations continue

to grow and encroach upon animal habitats, what are the possible effects on future EIDs?

2. Compare and contrast the three pathways by which a pathogen can "jump" from animal to human hosts: direct zoonotic transmission, intermediate host followed by zoonotic transmission, and cold/food chain transmission. How does transmission occur in each circumstance? What known diseases have been traced to each pathway?

3. What is the history of human-created pathogens infecting populations? By what methods have lab-created EIDs been introduced to human hosts, and what has been the effect? How have governments responded to such incidents?

4. Investigate the way in which China has shared or withheld data about the origins of SARS-CoV-2 with the international community. What information did researchers and governments seek from China? What information was shared and when? What information have researchers been unable to obtain from China? What factors may have affected China's decisions about sharing data? How has China's approach to collaborating with the international community affected the debate about competing theories regarding the origins of the virus?

5. Examine the role of the correlation fallacy *cum hoc ergo propter hoc* ("with this, therefore because of this") in the promulgation of laboratory leak theory. When Senator Tom Cotton suggested in February 2020 that the research laboratory in Wuhan could be the source of SARS-CoV-2, what evidence existed to support this claim? Was there any evidence to support the theory that research at the laboratory actually caused the virus to arise? How did the correlation between the outbreak of the virus in Wuhan and the presence of a research laboratory in Wuhan affect public perception of the origins of the virus?

6. In what ways do partisan political factors shape both the public perception of the origins of EIDs and the scientific community's ability to discover definitive answers?

7. How have governments and populations in the United States and European nations reacted to past EIDs that arose in Asia, Africa, and South America? How have policies and public discourse in Western nations affected the world's understanding of how such EIDs came to be?

NOTES

1. CDC Fact Sheet "Severe Acute Respiratory Syndrome," January 13, 2004.
2. WHO Meeting Report "Summary of probable SARS cases with onset of illness November 1, 2002 to July 31, 2003 (based on data as of December 31, 2003).
3. Wendong Li, Zhengli Shi, et al. "Bats are natural reservoirs of SARS-like coronaviruses." *Science*, October 28, 2005, volume 310, pages 676–679, 10.1126/science.1118391.
4. Peng Zhou, et al. "A pneumonia outbreak associated with a new coronavirus of probable bat origin." *Nature*, March 12, 2020, volume 579, pages 270–273, published online February 3, 2020, https://doi.org/10.1038/s41586-020-2012-7.
5. Hswen Y et al. Association of "#COVID19" versus "#Chinesevirus" with anti-Asian sentiments on Twitter: March 9–23, 2020. *Am J Public Health*, 2021, 111(5), pages 956–964. Doi:https//doi.org/10.2105/AJPH.2021.306154.
6. Pekar, J et al., "Timing the SARS-CoV-2 index case in Hubei province," *Science*, April 23, 2021, volume 372, pages 412–417.
7. "COVID-19—China," World Health Organization Disease Outbreak News, January 5, 2020, www.who.int/emergencies/disease-outbreak-news/item/2020-DON229, accessed August 18, 2021.
8. "COVID-19—China," World Health Organization Disease Outbreak News, January 12, 2020, www.who.int/emergencies/disease-outbreak-news/item/2020-DON233 accessed August 18, 2021.
9. Morens, DM et al., "The challenge of emerging and re-emerging infectious disease," *Nature*, July 8, 2004, volume 430, pages 242–249, doi:10.1038/nature08554.
10. Centers for Disease Control and Prevention, National Center for Emerging and Zoonotic Infectious Diseases.
11. Ghai, RR, et al., "Animal reservoirs and hosts for emerging alphacoronaviruses and betacoronaviruses," *Emerging Infectious Disease*, April 4, 2021, volume 27, pages 1015–1022, doi: 10.3201/eid2704.203945.
12. Jones, KE, et al., "Global trends in emerging infectious disease," *Nature*, February 21, 2008, volume 451, pages 990–994, doi:10.1038/nature06536.
13. Haider, N et al., "COVID-19—zoonosis or emerging infectious disease?" *Frontiers in Public Health*, November 26, 2020, volume 8, article 596944, pages 1–8, doi: 10.3389/fpubh.2020.596944.
14. United Nations Environment Programme and International Livestock Research Institute (2020). Preventing the next pandemic: zoonotic diseases and how to break the chain of transmission, Nairobi, Kenya, https://wedocs.unep.org/bitstream/handle/20.500.11822/32316/ZP.pdf?sequence=1&isAllowed=y.
15. Plowright, RK, et al., "Pathways to zoonotic spillover," *Nature*, August, 2017, volume 15, page 502–510, doi: 10.1038/nrmicro.2017.45, published online May 30, 2017.
16. Velasco-Villa, A et al., "The history of rabies in the Western Hemisphere," *Antiviral Research*, October 2017, volume 146, pages 221–232, doi: 10.1016/j.antiviral.2017.03.013.
17. Xiao, X, et al, "Animal sales from Wuhan wet markets immediately prior to the COVID-19 pandemic," *Nature*, June 7, 2021, volume 11, number 11898, doi: 10.1038/s41598-021-91470-2.

18. Brook, CE and Dobson, AP, "Bats as 'special' reservoirs for emerging zoonotic pathogens," *Trends in Microbiology*, March 2015, volume 23, number 3, pages 172–180, http://dx.doi.org/10.1016/j.tim.2014.12.004.

19. In his March 11, 2020 media briefing, WHO Director-General stated "we have therefore made the assessment that COVID-19 can be characterized as a pandemic" formally making these remarks for the first time.

20. Cotton made various speculative claims about the origins of SARS-CoV-2 during an interview on Fox News's "Sunday Morning Futures" on February 16, 2020.

21. The original February 17, 2020 Washington Post story "Tom Cotton keeps repeating a coronavirus *conspiracy theory* that scientists have disputed" was changed in a Washington Post correction to the phrase "fringe theory" on May 28, 2021 after President Biden directed the "Intelligence Community" to sort out SARS-CoV-2's origins.

22. The National Institutes of Health defines "gain-of-function" (GOF) as a "type of research that modifies a biological agent so that it confers new or enhanced activity to that agent…the subset of GOF research that is anticipated to enhance the transmissibility and/or virulence of potential pandemic pathogens, which are likely to make them more dangerous to humans, has been the subject of substantial scrutiny and deliberation," www.nih.gov/news-events/gain-function-research-involving -potential-pandemic-pathogens.

23. "an artificial origin of SARS-CoV-2 is not a baseless conspiracy theory that is to be condemned" from Segreto, R., & Deigin, Y. (2020). "The genetic structure of SARS-CoV-2 does not rule out a laboratory origin." *BioEssays*, e2000240. https://doi.org /10.1002/bies.202000240.

24. Bloom, JD, et al., "Investigate the origins of COVID-19," *Science*, May 14, 2021, volume 372, issue 6543, page 694.

25. Holmes, EC, et al, "The origins of SARS-CoV-2: a critical review," *Cell* volume 184, September 16, 2021, pages 4848–4856.

26. Humans and chimps also have about a 4% genomic sequence gap as first described in 2005 by the Chimpanzee Sequencing and Analysis Consortium in their landmark work "Initial sequence of the chimpanzee genome and comparison with the human genome," *Nature*, September 1, 2005, volume 437, pages 69–87, https:// doi.10.1038/nature04072.

27. Statement by President Joe Biden on the Investigation into the Origins of COVID-19, May 26, 2021.

3

COVID-19 Disease Manifestations

The Field: The symptoms of infectious disease—whether a fever, a cough, or a skin lesion—are the first clues that a person is ill. But at that time, the disease has already passed through one phase of its existence, and there will be more to come. Every infectious disease has clinical stages in which different symptoms appear. Underlying these is the pathophysiology that gives rise to the manifestations of disease in an infected person. Many aspects of COVID-19, the disease caused by the SARS-CoV-2 virus, have become clear, especially symptoms and clinical stages. Ongoing research into its pathophysiology continues to reveal the mechanisms behind these disease manifestations.

Field Sightings: primary ("chancre") phase, secondary ("rash") phase, tertiary ("organ failure") phase, symptoms, asymptomatic stage of illness, mild stage of illness, moderate stage of illness, severe stage of illness, critical stage of illness, host factors, comorbid conditions, pathophysiology, angiotensin converting enzyme 2 (ACE2), transmembrane protease serine S2 (TMPRSS2), pulmonary alveoli, alveolus-capillary interface, Type I alveolar epithelial cells (AEC1), Type II alveolar epithelial cells (AEC2), pulmonary surfactant, acute respiratory distress syndrome (ARDS), post-acute sequelae of COVID-19 (PASC), multisystem inflammatory syndrome in children (MIS-C).

FIELD EXPEDITION: THE THREE PHASES OF SYPHILIS

Syphilis is a sexually transmitted disease caused by the spiral-shaped bacterium *Treponema pallidum*, called a "spirochete" for its characteristic

DOI: 10.1201/9781003310525-4

appearance. An infectious disease with an infamous history spanning centuries, syphilis remained incurable until the discovery of penicillin by Alexander Fleming in 1928, and its widespread use against syphilis began in 1943, leading to a precipitous drop in cases.

However, before the advent of penicillin, people with untreated syphilis lived through the clinical course of this disease from initial infection to eventual death, allowing its natural progression to be observed such that now syphilis is known by three well-recognized phases.

The ***primary ("chancre") phase*** of syphilis occurs 10 to 90 days after someone is exposed to the syphilis spirochete through sexual contact. The infected person develops a small, firm, painless skin lesion called a "chancre" on the penis, vulva, mouth, or anus. This chancre is often not detected due to its painlessness and small size, disappearing within weeks without leaving a trace.

The ***secondary ("rash") phase*** of syphilis occurs two weeks to six months after the chancre fades. This phase is dominated by an astonishing generalized rash over the entire body (see Figure 3.1) Left in this untreated

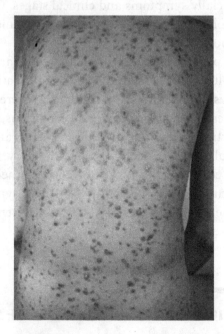

FIGURE 3.1
Generalized rash of secondary syphilis on upper torso and extremities. (Photo credit: CDC/Dr. Gavin Hart.)

state, secondary syphilis enters a latency period of some two years before entering its next phase of illness.

The **tertiary ("organ failure") phase** of syphilis may not appear for several decades after initial exposure. This lethal phase is known for widespread internal organ damage of the cardiovascular system with aneurysms, coronary artery disease, and congestive heart failure; "gummatous syphilis" with large destructive skin and bone lesions; and "general paralysis of the insane," an antiquated name for the spinal cord damage and psychotic illness associated with late-stage syphilis.

LESSONS LEARNED: THE THREE PHASES OF SYPHILIS

Syphilis has plagued humankind for centuries, and effective treatment for it has existed for only a fraction of that time, giving scientists ample opportunity to observe the natural progression of the disease in the human body. While all diseases have characteristic clinical stages, not all present such a clearly delineated set of specific, identifiable phases as syphilis. The structured nature of the disease's progression allows doctors to clearly identify symptoms, assess damage, and prescribe appropriate antibiotics that can halt disease progression and, in most cases, cure an illness that once carried a prognosis of near-certain death.

Like syphilis, COVID-19 progresses through distinct phases characterized by mild to moderate illness which may progress to a serious, life-threatening stage ending either in death or, for some, lingering symptoms for months or years. COVID-19 is a very new infectious disease whose pathophysiology and clinical course is still being revealed, but the highly structured nature of the clinical stages provides healthcare professionals clarity in treating symptoms and anticipating—and perhaps preventing—additional complications.

CONCEPTS COVERED IN THIS CHAPTER

- Symptoms
- Clinical stages

- Host factors
- Comorbid conditions
- COVID-19 pathophysiology

COVID-19 Is a New Disease Entity

COVID-19 is an emerging infectious disease (EID)—a brand-new pathogen just 30 months old. Its causative agent (the SARS-CoV-2 virus) and route of transmission (airborne) are known, but many aspects of the disease are still being determined. Even in its short lifetime, however, much has been learned about COVID-19. It was initially viewed as a respiratory illness but is now understood to be a multisystem disorder that moves through distinct phases, spurred on by a dysregulated immune system and prominently featuring blood coagulation disorders (see Figure 3.2).

All infectious diseases are distinguished by symptoms experienced by those who are ill, clinical stages through which a disease progresses over time, host factors and comorbid conditions which increase an individual's susceptibility to illness, and an explanatory pathophysiology underlining the disordered mechanisms causing the disease.

Symptoms

An illness usually begins with *symptoms* such as fever, body aches, cough, tiredness, or stomach upset. If symptoms persist or worsen, an ill person will usually seek help from medical professionals.

Physicians and nurse practitioners are trained to evaluate the cause of symptoms through detailed questioning, physical examination, and use of laboratory tests, imaging studies, or other diagnostic modalities which may yield a diagnosis. Once a diagnosis is made, interventions with surgery, medications, or other agents can be employed to relieve symptoms and, hopefully, cure the ailment, allowing restoration of normal function.

COVID-19 Symptoms

SARS-CoV-2 is transmitted by respiratory means; after exposure, the incubation period for developing active infection may be up to 14 days, with a median of four to five days. Once inhaled through the nose or mouth, SARS-CoV-2 infects nasal epithelial cells, producing loss of smell

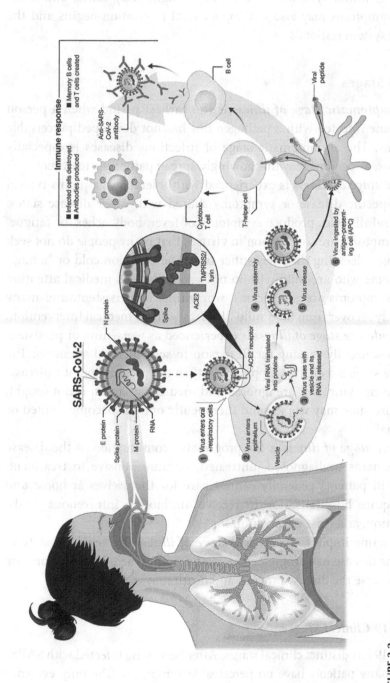

FIGURE 3.2

Transmission and life cycle of SARS-CoV-2, causing COVID-19. (Source: Ian Dennis/Wikimedia Commons/licensed under CC BY 4.0.)

or taste as initial symptoms. Fever, dry cough, body aches, and other flu-like symptoms may also develop as viral replication begins and the immune system responds.

Clinical Stages

The *asymptomatic stage of illness* is the earliest period when a person has become infected with a pathogen but has not developed perceivable symptoms. The asymptomatic stage of infectious diseases is especially dangerous, as persons may unknowingly spread pathogens to others.

A *mild stage of illness* is experienced with tolerable symptoms typical of that specific disease or symptoms shared with other disease states. Many viral illnesses produce symptoms of fever, body aches, or fatigue. These symptoms are so common to viruses that many people do not seek a diagnosis, deeming that they suffer from the common cold or "a bug." Most people who are mildly ill do not generally seek medical attention if their symptoms are not serious, worsening, or persistent, and many completely recover from their mild illnesses with no medical intervention.

The *moderate stage of illness* is experienced as worsening or persistent symptoms, usually causing the ill person to seek medical attention. The moderate stage is a crucial "tipping point" in the progression of a disease. If worsening symptoms are ignored and medical attention is not sought, the disease state may worsen and therapeutic options become limited or ineffectual.

A *severe stage of illness* is the progressive continuation of the disease which remains undiagnosed, untreated, or unresponsive to treatment. Severely ill patients generally cannot care for themselves at home and often require hospitalization to receive antibiotics, intravenous fluids, oxygenation, and close observation.

As its name implies, the *critical stage of illness* is a terminal state in which the disease has progressed beyond a point where interventions are able to reverse the disease process, and death may occur.

COVID-19 Clinical Stages

COVID-19 has distinct clinical stages. After becoming infected with SARS-CoV-2, many patients have no perceivable symptoms. The only evidence of illness in these asymptomatic persons is a positive antigen or nucleic acid amplification test (NAAT) like the commonly used real-time reverse transcriptase polymerase chain reaction test (rRT-PCR or PCR test).

Most SARS-CoV-2 infected people progress to a mild or moderate stage of illness, especially those who are generally healthy or become infected after being vaccinated against COVID-19. Symptoms of fever, sore throat, dry cough, muscle aches, and general tiredness are commonly experienced. Shortness of breath or productive cough may indicate a worsening disease course.

Some COVID-19 patients progress to severe or critical stages, particularly unvaccinated persons, older people, or anyone with chronic medical or immunocompromising conditions.

Host Factors

The pathophysiology of a disease is often determined by *host factors*, which are the physical characteristics of persons that may predispose them to affliction or cause them to experience more serious clinical stages of illness.

Age, sex, and socioeconomic status are examples of host factors. Older persons, for example, have less robust immune systems, predisposing them to worse disease outcomes. Some diseases tend to affect people of different sexes more or less severely. People of lower socioeconomic status may lack health insurance or access to healthcare professionals or facilities, limiting their ability to seek medical care or receive treatment.

COVID-19 Host Factors

Advanced age, male sex, and lower socioeconomic status either increase susceptibility to being infected with SARS-CoV-2 (through occupational exposure) or predict progression to more serious stages (due to "immunosenescence," or a less robust immune system with advancing age).

Comorbid Conditions

Similar to host factors, *comorbid conditions*—pre-existing health conditions like diabetes, hypertension, or immunocompromising conditions—may predispose individuals to becoming afflicted with an ailment or experiencing more serious outcomes.

COVID-19 Comorbid Conditions

Elevated blood glucose in people with diabetes damages the internal endothelial lining of blood vessels, setting the stage for atherosclerosis, which is further stressed by thrombotic events associated with COVID-19. People taking immunosuppressing medications are less able to develop an appropriate immune reaction in response to SARS-CoV-2 infection.

Pathophysiology

Clinical stages are the perceived states experienced by patients as symptoms and categorized by medical professionals as diagnoses. The clinical stages of an illness are outward facing: perceptible by patients, observed by health professionals.

What drives a disease and determines its clinical progression from mild to severe is inward facing, defined by its *pathophysiology*, or the disordered physiological processes giving rise to a disease state. Understanding the pathophysiology of a disease predicts the natural progression of the disease from stage to stage, allowing time to prepare, plan, and institute appropriate interventions.

COVID-19 Pathophysiology

SARS-CoV-2 uses the *angiotensin converting enzyme 2 (ACE2)* receptor to enter host cells, assisted by the priming actions of host cell membrane proteases like *transmembrane protease serine S2 (TMPRSS2)*. ACE2 receptors are found all along the respiratory tract, from nasal epithelial cells to the alveolar sacs where oxygen–carbon dioxide gas exchange occurs, but ACE2 receptors are also prominently found in gastrointestinal, renal, cardiac, and vascular cells, which may explain the signs, symptoms, and complications that arise from SARS-CoV-2 infection.

ACE2 Receptors, Fluid Balance, and Inflammation

ACE2 receptors are a modulating component of the renin-angiotensin-aldosterone system (RAAS), among the most important of the hormone-mediated physiologic systems in humans. RAAS regulates blood pressure and fluid balance through renin, angiotensin II, and aldosterone.

When activated, angiotensin II has powerful vasoconstrictive proinflammatory actions. ACE2 prevents angiotensin II activation; when actions of ACE2 are blocked by SARS-CoV-2, the unfettered actions of angiotensin II are unleashed, contributing to inflammation and acute respiratory distress syndrome.

Alveolar Epithelial Cells

Though SARS-CoV-2 first enters nasal epithelial cells, the virus may travel to the furthest point in the respiratory tract: the *pulmonary alveoli* in the lungs, which are small sacs richly enmeshed in blood vessels where inhaled oxygen is exchanged with carbon dioxide in the bloodstream. Oxygen molecules cross the *alveolus-capillary interface* and are captured by hemoglobin molecules within circulating red blood cells in exchange for carbon dioxide gas, which fills the alveolar space and is exhaled through the lungs.

The interior passages of the windpipe (trachea), air ducts (bronchi), and air sacs (alveoli) are lined by specialized epithelial cells. The majority of epithelial cells lining alveoli are *Type I alveolar epithelial cells (AEC1)*, which allow oxygen–carbon dioxide gas exchange and maintain the structural integrity of the alveolus-capillary interface.

Though much smaller in quantity, *Type II alveolar epithelial cells (AEC2)* play three major roles in respiratory physiology: AEC2 cells produce *pulmonary surfactant*, a substance without which gas exchange would not be possible; AEC2 cells are the stem cells which replenish AEC1 cells; and AEC2 cells participate in immune defense of the alveolar space.

AEC2 cells are rich in the angiotensin converting enzyme 2 receptor (ACE2), the entry point for SARS-CoV-2. ACE2 receptors promote an anti-inflammatory milieu in the cells and tissues where they are found. When they are "downregulated" or made inactive by SARS-CoV-2 virus receptor blockade, an immune system inflammatory response ensues.

Acute Respiratory Distress Syndrome

Acute respiratory distress syndrome (ARDS) was the leading cause of death for COVID-19 patients during the initial months of the pandemic. ARDS is a life-threatening complication resulting from pneumonias of any cause, severe burns, and trauma. ARDS causes a severe drop in blood oxygenation levels (hypoxemia), fluid accumulation in the lungs (pulmonary edema), and "stiff lungs" noted during mechanical ventilation. SARS-CoV-2 causes damage to AEC1 cells, which allows vascular fluid to pass into alveolar spaces, and damage to AEC2 cells, which decreases production of pulmonary surfactant: these are the leading causes of ARDS pathophysiology.

PASC and MIS-C

After being discharged from hospitalization or recovering at home from COVID-19, some patients may continue to experience lingering symptoms of lethargy, body aches, and mental incoherence commonly called "brain fog." This is particularly prevalent in post-discharge patients, but even those experiencing moderate COVID-19 have reported these lingering symptoms called "long COVID" or *post-acute sequelae of COVID-19 (PASC)*. Models for predicting patients likely to develop PASC, specific diagnostic studies, and interventions are being studied, but lingering symptoms after serious viral illness are not uncommon and have been observed in patients recovered from Ebola virus.

Children are unlikely to develop serious complications of COVID-19, but some rare cases of *multisystem inflammatory syndrome in children (MIS-C)* with fever, hypotension, and multiple organ involvement have been reported in children with COVID-19.

Looking Back, Looking Ahead

Every new disease is an initial mystery: scientists must discover the causative agent, symptoms, clinical stages, host factors, comorbid conditions, and, ultimately, the pathophysiology that explains the mechanisms that allow a pathogen to cause disease in a human host.

There is still much to be learned about COVID-19 and the SARS-CoV-2 virus that causes it, but enough has been discovered to enable effective vaccines and therapeutic agents to be developed. Importantly, the asymptomatic and mild stages of the disease are fairly well understood, which gives medical professionals a better chance of intervening before illness progresses to more advanced stages.

Compared to bacteria-induced disease, viruses can be particularly hard to treat; antiviral therapeutics are notoriously difficult to develop. While the great promise of Alexander Fleming's discovery of penicillin in 1928 was the ability to effectively target many bacteria that cause disease and curtail centuries of suffering caused by diseases like syphilis, no such medication yet exists for viruses. Ongoing research into COVID-19 will reveal more about the full breadth of its disease manifestations, and this knowledge will benefit those suffering from COVID-19 as well as victims of future EIDs.

FUTURE OUTINGS

To understand how the lessons of COVID-19 can be applied to future EIDs, ongoing excursions into the field must explore:

- The pathophysiology of COVID-19, especially the factors that lead to "long COVID" and the severe and critical stages of illness
- The host factors that increase susceptibility to COVID-19, particularly the effects of economic inequality and oppression of marginalized populations
- Variations in transmissibility, symptoms, and pathophysiology caused by COVID-19 variants

Student Research Questions

1. What viruses cause common symptoms of mild illnesses such as fever, body aches, or fatigue? Why do these illnesses infrequently progress to severe clinical stages? How does their pathophysiology compare to viruses that often cause more serious illness, such as COVID-19?
2. At what point in the progression of the pandemic did researchers come to understand that COVID-19 was not simply a respiratory illness? What factors led to a more complete knowledge of the symptoms and pathophysiology of the disease?
3. What aspects of COVID-19 pathophysiology are best understood and least understood? What avenues of research are being pursued?
4. Compared to other EIDs that arose in the twenty-first century, how much is known about COVID-19 pathophysiology? How

does knowledge about other coronavirus-induced diseases inform understanding of and research into COVID-19?

5. What other diseases display clearly structured clinical stages like syphilis and COVID-19? How do medical professionals use knowledge of these stages to treat diseases?

4

Testing

The Field: When a person experiences symptoms of illness, they look to healthcare practitioners to tell them why they're sick and provide a remedy. But what of people who have an infection but show no symptoms—how can they, and the people around them, know if they are spreading disease? In a global viral pandemic, this question is of utmost importance. Clinical tests are one component of making an accurate diagnosis. Not all tests provide the same level of accuracy, and knowing how to interpret test results, in conjunction with other diagnostic criteria, is paramount for healthcare workers and public policymakers. The three very different kinds of SARS-CoV-2 tests available offer varying pathways to identifying infected people and implementing testing programs.

Field Sightings: clinical tests, diagnostic test, screening test, sensitivity, specificity, true positive result, false negative result, true negative result, false positive result, "gold standard" or reference standard test, positive predictive value (PPV), negative predictive value (NPV), Bayes's theorem of pre-test probability, prevalence, differential diagnosis, incubation period, serologic antibody tests, community seroprevalence studies, quantitative real-time polymerase chain reaction tests (qRT-PCR or PCR), Taq DNA polymerase, complementary DNA (cDNA), cycle threshold value (Ct), viral load, replication-incompetent, time-based or symptom-based approach, antigen rapid diagnostic tests (Ag-RDTs or RDTs), lateral flow immunoassays (LFIAs), turnaround time (TAT).

DOI: 10.1201/9781003310525-5

FIELD EXPEDITION: MICHAEL MINA'S MISSION

Like every doctor and scientist in 2020, Michael Mina, MD, PhD, wanted to stop the spread of SARS-CoV-2 as quickly as possible. His laser focus, though, was on testing—specifically, getting as many antigen rapid diagnostic tests into the hands of as many people as possible.

As a research scientist trained in medicine and epidemiology, Mina is interested in the disease experience of both individual patients and whole populations. During pandemics, diagnosing one acutely sick person may be as important as diagnosing disease among many people who show no signs of illness. As SARS-CoV-2-infected people in the asymptomatic phase of disease—which can last as long as 14 days—moved through the world, they were unknowingly transmitting the virus to others. During the early stages of a mounting viral pandemic in which asymptomatic persons were spreading new infections, Mina argued that identifying and isolating these people as quickly as possible was of paramount importance.[1]

Within six months of the World Health Organization (WHO) declaring COVID-19 a global pandemic, two SARS-CoV-2 diagnostic tests were in widespread use: the "gold standard" PCR (quantitative real-time polymerase chain reaction test) and the "good enough" RDT (antigen rapid diagnostic test). Mina was critical of how the PCR test had been elevated over RDTs. While PCR tests are superior in accurately diagnosing the presence of SARS-CoV-2 virus, they are performed by trained personnel in clinical laboratories, which is not only costly but, worst of all for Mina, takes too much time.

Mina's position held that in a general population, the deployment of inexpensive, easily administered RDTs whose results could be known within minutes would do more to control the spread of SARS-CoV-2 than costly, lab-performed PCR tests whose results would not be known for days, even if the timelier RDTs were less sensitive than PCR tests.

Mina made similar arguments in a retrospective cohort study of antigen test use among healthcare workers in a teaching medical center, observing, "Ag-RDT may be a better surrogate for infectivity than PCR, as it represents translated viral proteins rather than RNA remnants."[2]

Mina made it his mission to promote the use of the "good enough" RDT in widespread testing regimens as a strategy to contain infection as early as possible.

LESSON LEARNED: MICHAEL MINA'S MISSION

Clinical tests are an important component of medical care and public health, useful in making the correct diagnosis of an illness in an individual patient or in identifying asymptomatic persons within a population during an infectious epidemic. These tests have trade-offs in terms of cost, ease of use, and timeliness of results.

The PCR test may be the "gold standard" in terms of accuracy, but if the cost and long wait for results allowed infected people to circulate among the population, could it really be considered the "gold standard" in containing the spread of disease? A pandemic demands that public health officials shift their thinking about testing and diagnosis; the best test for an individual may not be the most effective option when looking at an entire population.

Ultimately, both the PCR and RDT tests have an important role to play in diagnosing COVID-19, but the strategy behind their use determines how effective they can be in the ultimate aim of containing the spread of infection.

CONCEPTS COVERED IN THIS CHAPTER

- Types of clinical tests
- Performance characteristics of clinical tests
- The process of diagnosing disease
- Factors in diagnosing COVID-19
- SARS-CoV-2 clinical tests

Types of Clinical Tests

Clinical tests are laboratory or radiology procedures used for medical decision making in individual patients or for public health purposes among whole populations. Healthcare providers interpret the data produced by a clinical test to determine if illness is present in a patient.

When a clinical test is used to confirm a suspected diagnosis in an ill person being evaluated by a healthcare professional, it is called a *diagnostic test*. When a clinical test is used to detect the presence of an illness or condition in a healthy person it is called a *screening test*. Diagnostic and screening tests are ordered by healthcare professionals[3] or public health practitioners for different reasons, and screening tests often detect the presence of a condition which needs to be confirmed by diagnostic tests.

While clinical tests are powerful decision aids, two factors must be considered when these tests are used in real-world settings: the analytic performance characteristics of the tests themselves and how tests results are interpreted in conjunction with the characteristics of individual patients and populations.

Performance Characteristics of Clinical Tests

Clinical tests vary in their ability to accurately detect the presence or absence of a condition. The ability of a test to state "this person truly has this condition" is called its *sensitivity*; conversely, the ability of a test to state "this person truly does *not* have this condition" is called its *specificity*.

Sensitivity and specificity describe the performance ability of a test to produce an accurate result. Based on observed results of the test in experimental settings, every test has a known sensitivity and specificity, which are expressed as percentages. The higher a test's percentage for sensitivity and specificity, the more accurate the results will be.

A test result may be positive or negative. A *true positive result* means the condition is truly present in the tested person. A *false negative result* means that the condition is truly present in the tested person, but the test falsely signals the condition is not present. Both true positive and false negative results are associated with the sensitivity of a test.

The specificity of a test indicates that a condition does not exist in the person being tested. In a test with high specificity, negative test results mean the tested person very likely does not have the condition; this is called a *true negative result*. If a positive result occurs when using a test with high specificity, the positive result is a *false positive result* because the condition is truly not present though the test result indicates otherwise. Table 4.1 indicates associations among sensitivity, specificity, and test results.

A clinical test with high sensitivity will produce many true positive results. But high sensitivity tests may also produce a small number of

TABLE 4.1

Associations Among Sensitivity, Specificity, and Test Results

	Has condition	Does not have condition	
Positive test result	True positive (TP)	False positive (FP)	Positive predictive value (PPV)
Negative test result	False negative (FN)	True negative (TN)	Negative predictive value (NPV)
	Sensitivity	Specificity	

false negative results. The sensitivity of a test is therefore determined by calculating TP/(TP + FN) and expressed as a percentage.

A clinical test with high specificity will produce many true negative results. But high specificity tests may also produce a small number of false positive results. The specificity of a test is therefore determined by calculating TN/(TN + FP) and expressed as a percentage.

Every clinical test has an associated sensitivity and specificity, and while no test is both 100% sensitive and 100% specific, clinical tests which come closest to these performance goals are called "gold standard," or reference standard, tests.

"Gold Standard" or Reference Standard Test

A clinical test which almost always correctly identifies a disease as being present or absent is called a *"gold standard" or reference standard test* for that condition and has very high sensitivity and specificity associated with its performance. Using Table 4.2, consider the following performance results for hypothetical Test A.

Using the previously described formula for calculating sensitivity and specificity, the performance data in this example indicates the sensitivity

TABLE 4.2

Hypothetical Test A Performance Results

	Has condition	Does not have condition
Positive test result	82	3
Negative test result	18	97
	Sensitivity	Specificity

of Test A as 82/(82+18), or 82%, and the specificity of Test A as 97/(97+3), or 97%.

Any test having 82% sensitivity and 97% specificity would be considered the reference standard test for the condition of interest, and this "gold standard" test would be the reference against which all other tests should be compared.

Because of the high sensitivity and specificity of reference standard tests, two mnemonics are helpful when interpreting results: SNOUT and SPIN. A highly sensitive test, if negative, "rules out" the condition—indicating the condition is truly not present (sensitive, negative, rules out, or SNOUT). A highly specific test, if positive, "rules in" the condition—indicating the condition is truly present in the tested individual (specific, positive, rules in, or SPIN).

Positive and Negative Predictive Values

Sensitivity and specificity describe the performance characteristics of the test itself. In real-world settings of clinical practice and public health, the pressing concern is to tell a patient what the results of a test actually mean. When a patient "tests positive" for infection, it may be a true or false positive result. Similarly, a "negative" test may be a true or false negative result. False positives and false negatives are inevitable with any clinical test, no matter how reliable. *Positive predictive value (PPV)* and *negative predictive value (NPV)* describe the proportions of true positive and true negative results, respectively, that a given clinical test will return.

PPV will include a certain number of false positive test results, which means some people will be over-diagnosed, i.e., told they have a condition when one does not exist. This can lead to anxiety and possible harm from unnecessary treatments. Similarly, NPV will include some false negative results, leading to some patients being told they do not have a condition when, in fact, they do. This leads to conditions remaining untreated and, in the case of infectious diseases, further spread of illness. These possible harms should be considered when evaluating the PPV and NPV of a test.

Clinical Test Results and Disease Probability

Deciding whether to conduct clinical tests and interpreting the data from any tests performed is only one part of making a diagnosis and subsequent treatment decisions. The use of prostate-specific antigen (PSA) tests to screen for prostate cancer is a prime example of the nuanced and complex way in which clinical tests are used in real-world scenarios.

Bayes's Theorem of Pre-Test Probability

In discussing sensitivity, specificity, reference standards, and predictive values, one is dealing with matters of chance or probability. Different clinical tests have different analytic performance characteristics, and before any test is performed one should have a very good idea about how to interpret a positive or negative result and what to do next with this information.

Thomas Bayes was an eighteenth-century mathematician whose theories about probability have been applied to clinical decision making. *Bayes's theorem of pre-test probability* holds that observing a phenomenon is unlikely when occurrence of that phenomenon in the population being observed is low.

Bayes's theorem indicates skepticism should be used in interpreting clinical test results when a test is performed on individuals who are unlikely to have the condition before the test is run or tests are performed in communities where the *prevalence*, or rate of occurrence, of a condition in a community is very low.

Making the Correct Diagnosis of a Disease

Many diseases share the same unspecified signs and symptoms. Since patients are not able to self-diagnose their afflictions, they rely on healthcare professionals to determine the reason they are sick. Making the correct diagnosis of a disease is not an academic exercise. An accurate diagnosis enables appropriate therapeutic interventions that can halt disease progression and hopefully produce a complete cure.

When healthcare professionals meet acutely ill people, their disease states are not initially known. No patient ever begins a conversation with their physician by saying "Doctor, I'm here today because I have right lower lobe pneumonia." Instead, the patient might say, "Doctor, I'm here today because I've got pain on the right side of my chest and a deep cough with sputum, which makes the pain even worse."

Upon hearing these complaints, the healthcare professional asks more questions (takes a medical history), touches or listens to different parts of the patient's body (performs a physical examination), and may schedule blood, urine, or radiology imaging (orders clinical tests).

At this point in the doctor-patient encounter, an answer to the patient's question of "Why am I sick?" may not be known. But after taking a medical history and performing a physical examination, healthcare professionals begin to develop a list of possible reasons explaining the patient's complaints, which is called the ***differential diagnosis***.

The differential diagnosis Is a list of possible reasons connecting the patient's symptoms, history, and physical exam to one or more disease states, from most likely to least likely. Seasoned healthcare professionals can often make a correct diagnosis purely based on the patient's history and physical examination alone. This is because disease presentations are similar from patient to patient, and providers with years of experience can quickly make an accurate diagnosis by merely hearing the patient's story or observing physical changes in the patient's body.

But even with a high degree of confidence in a diagnosis based on medical history and physical examination, definitive disease confirmation via a clinical test is preferred before starting specific therapeutic interventions.

Using Clinical Tests in Making Diagnoses

Diagnostic tests are confirmatory tests used in conjunction with medical histories and physical examinations to "rule in" or "rule out" specific diseases that are part of the provider's differential diagnosis. Experienced healthcare providers generally have one or two leading diagnostic theories explaining the story presented by the patient; providers select a series of complementary tests to provide additional evidence that a specific diagnosis explains the patient's disease state.

In the example of the patient with chest pain and cough above, the medical history might indicate the patient had recently taken a vacation and stayed in a cabin with old plumbing and had taken long hot showers, breathing in steam. Three days later, after returning home, he felt vaguely ill and developed right-sided chest pain made worse by breathing or coughing episodes with copious amount of sputum. On physical examination, the healthcare provider noted that the patient leaned to his right side to "baby" his sore chest and had a sweaty brow. His temperature was 101 °F and he had raspy loud noises, called "rhonchi," when his lungs were auscultated with a stethoscope.

Based on the patient's history of breathing water droplets, right-sided pleuritic chest pain, productive cough, fever, and lung rhonchi, the provider suspected Legionnaire's pneumonia and ordered a series of confirmatory diagnostic tests to "rule in" this diagnosis: pulse oximetry to get a general sense of how well the patient was oxygenated, chest x-ray to visualize the area(s) affected by pneumonia, sputum culture to determine the exact pathogen causing the pneumonia, and complete blood count to see if white blood cells were elevated as further indication of an active infection.

These complementary diagnostic tests contributed to the mounting evidence that the patient not only had bacterial pneumonia but a specific bacterial pneumonia caused by *Legionella pneumophila*: a disease known as Legionnaire's disease. The provider could then prescribe the most effective therapeutic intervention: a fluroquinolone or macrolide antibiotic.

Making the Correct Diagnosis of COVID-19

SARS-CoV-2 virus is transmitted through respiratory means to those within about six feet of an infected person. The virus is inhaled into the nostrils and windpipe, latches on to host cell ACE2 receptors and enters these cells, where it begins making copies of itself. The viral replication time when new copies of virus are being produced is called the *incubation period*; for SARS-CoV-2, this is usually five days but may be as long as 14 days.

During this incubation period, many SARS-CoV-2 infected people remain asymptomatic, unknowingly spreading the virus through continued close contact with others. During or shortly after this period, people may develop fever, headaches, tiredness, dry cough, or muscle aches. Because these are nonspecific symptoms common in many conditions, the differential diagnosis for COVID-19 may include the common cold, seasonal allergies, or influenza. Other symptoms may also develop, including loss of smell or taste and diarrhea. As these are not common with colds or allergies, they may specifically indicate SARS-CoV-2 infection.

In addition to observing symptoms, healthcare practitioners can determine host factors and comorbid conditions by taking a medical history to determine the likelihood of SARS-CoV-2 infection. The prevalence of circulating virus in the area the patient lives is another factor to consider.

Predisposing risk factors in symptomatic people and high prevalence of circulating virus increase the pre-test probability that clinical tests will likely show positive results. Clinical tests are essential for confirming COVID-19 diagnoses, as well as screening asymptomatic patients for SARS-CoV-2 infection.

SARS-CoV-2 Clinical Tests

There are three types of tests that can detect SARS-CoV-2 infection:

- Serologic antibody tests
- Quantitative real-time polymerase chain reaction tests (qRT-PCR or PCR)
- Antigen rapid diagnostic tests (Ag-RDTs or RDTs)

They all differ in their methodology and performance characteristics—as well as how they are used for screening and diagnostic testing.

Serologic Antibody Tests

Serologic antibody tests are blood tests used to detect the presence of immunoglobulins created sequentially during SARS-CoV-2 infection: immunoglobulin A (IgA), which is immediately made and then dissipates; immunoglobulin M (IgM), which forms and gradually disappears; and immunoglobulin G (IgG), which takes more time to develop but is longer lasting. IgA, IgM, or IgG might not be detected in people taking immunosuppressing drugs (for instance, organ transplant recipients) or those having immunocompromising conditions.

Because it may take up to three weeks after symptoms develop for detectable antibodies to be made, serologic tests are not used to diagnose SARS-CoV-2 infection (see Figure 4.1).

These tests take a backwards look at the infection stage and cannot be used to screen out or diagnose active SARS-CoV-2 infection. What serologic tests can do is determine the prevalence or spread of infection within a community. Serologic tests are therefore important parts of ***community seroprevalence studies***.

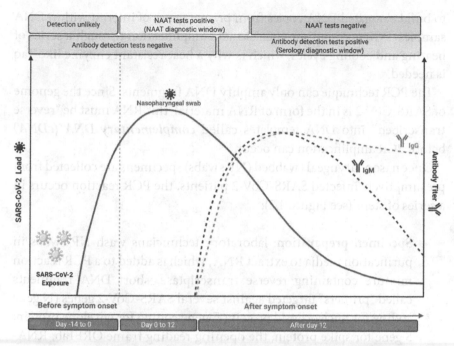

FIGURE 4.1
Antibody detection levels after SARS-CoV-2 infection. (Source: Sethuraman et al. This figure was elaborated from the data contained in the work of Sethuraman et al. [213]. Created with BioRender.com. © 2020 by the authors. Licensee MDPI, Basel, Switzerland.)

PCR Tests

Quantitative real-time polymerase chain reaction (qRT-PCR or PCR) tests use nasopharyngeal specimens to detect RNA genetic material from SARS-CoV-2 virus. They require complicated laboratory processing methods to detect SARS-CoV-2 genetic material, and a decision about whether SARS-CoV-2 virus is present may take hours.

The polymerase chain reaction test is a laboratory technique developed in 1985 by Kary B. Mullis,[4] which detects genes of interest from DNA samples. PCR allows millions of detectable DNA copies to be created from small quantities of DNA, a technique which revolutionized various disciplines—including genetic testing, AIDS research, criminology, and paleontology—and is now being most recently used to detect small amounts of RNA extruded by SARS-CoV-2 virus.

PCR uses a heat-resistant DNA polymerase enzyme called ***Taq DNA polymerase*** harvested from the heat-tolerant ***Thermus aquatics*** bacterium to build elongated DNA copies from primer genes of interest within DNA samples. PCR accomplishes nucleic acid amplification through a series of heating and cooling cycles, which is why a heat-resistant enzyme like Taq is needed.

The PCR technique can only amplify DNA fragments. Since the genome of SARS-CoV-2 is in the form of RNA material, this RNA must be "reverse transcribed" into DNA templates called *complementary DNA (cDNA)* before PCR amplification can occur.

Once nasopharyngeal swabbed (NP swabs) specimens are collected from presumptively infected SARS-CoV-2 patients, the PCR reaction occurs in a series of steps (see Figure 4.2):

- **Specimen preparation**: laboratory technicians wash NP swabs in purification media to extract RNA, which is added to a PCR reaction mixture containing reverse transcriptase, short DNA fragments called "primers" targeted against several SARS-CoV-2 genes (N gene coding for nucleocapsid protein; E gene coding for envelope protein; S gene for spike protein; the opening reading frame ORF1ab; RNA-dependent polymerase (RdRP) which is unique to RNA viruses); Taq DNA polymerase, and DNA nucleotides.
- **Heat denaturation (96°C)**: the PCR reaction mixture containing complementary DNA created by reverse transcriptase, which converts SARS-CoV-2 RNA genetic material into transcribable cDNA, is heated to a high temperature to "denature" the double-stranded DNA helix, exposing its separate strands.
- **Primer annealing (55–65°C)**: the reaction mixture is cooled, which promotes annealing of DNA primers targeted against the various SARS-CoV-2 gene regions described above to various cDNA stands, which happens with a high degree of specificity because so many SARS-CoV-2 gene regions are targeted as potential DNA primers.
- **Primer extension (72 °C)**: as the reaction mixture is reheated, Taq DNA polymerase begins to add available DNA nucleotides to the single stranded template space around the annealed DNA primer areas, extending and elongating newly forming double-helix DNA.
- **Repeat heating and cooling**: heat denaturation, primer annealing, and primer extension comprise one complete PCR cycle, which is then repeated through 40 cycles, exponentially increasing the amount of specific DNA. Fluorescent probes are also included in the

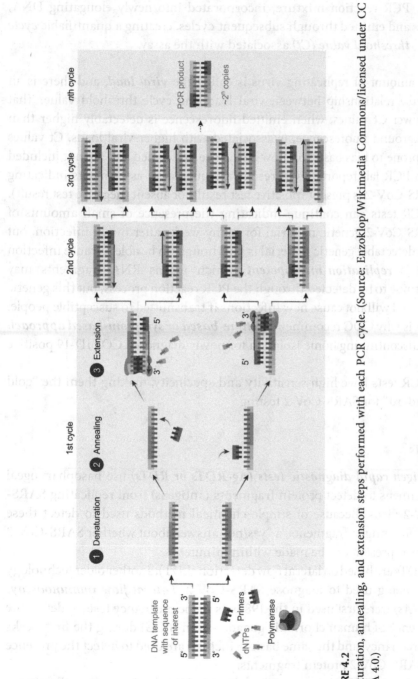

FIGURE 4.2

Denaturation, annealing, and extension steps performed with each PCR cycle. (Source: Enzoklop/Wikimedia Commons/licensed under CC BY-SA 4.0.)

PCR reaction mixture, incorporated into newly elongating DNA, and emitted through subsequent cycles, creating a quantifiable *cycle threshold value (Ct)* associated with the assay.

The amount of replicating virus is called the *viral load*, and there is an inverse relationship between viral load and cycle threshold value. That is, lower Ct values, when emitted fluorescence is detectably higher than background fluorescence, is associated with higher viral loads. Ct values are prone to bias based on how assays are performed and are not included with PCR lab reports. PCR results are qualitative assessments indicating SARS-CoV-2 is present (positive test result) or absent (negative test result).

PCR tests can continue indicating the presence of small amounts of SARS-CoV-2 genetic material for many weeks after initial infection, but this detectable genetic material is not thought to be able to cause infection and is *replication-incompetent*, which means RNA fragments may continue to be detected through the PCR reaction process, but this genetic material will not cause new infections if transmitted to susceptible people. This is why CDC recommends a *time-based or symptom-based approach* for discontinuing home isolation for newly diagnosed COVID-19 positive people.

PCR tests have high sensitivity and specificity, making them the "gold standard" for SARS-CoV-2 testing.

RDTs

Antigen rapid diagnostic tests (Ag-RDTs or RDTs) use nasopharyngeal specimens to detect protein fragments (antigens) from replicating SARS-CoV-2 virus. Because of simple chemical methods used to detect these protein antigen fragments, a "yes/no" answer about whether SARS-CoV-2 virus is present can be made within minutes.

RDTs are based on lateral flow detection (LDT), another older technology now being used to diagnose SARS-CoV-2. *Lateral flow immunoassays (LFIAs)* were first used in the 1980s as urine pregnancy tests to detect the presence of human chorionic gonadotropin (hCG) during the first weeks of pregnancy, and the same basic principles are used to detect the presence of SARS-CoV-2 protein fragments.

As with PCR testing, antigen-based RDT begins with collecting a nasal specimen by swabbing the internal surfaces of each nostril. For

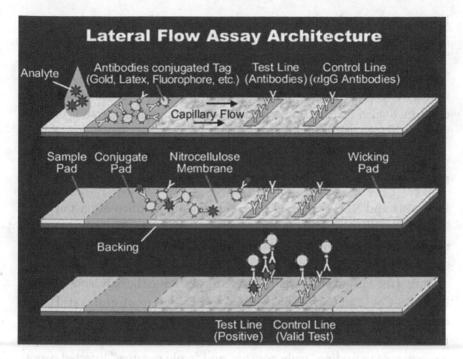

FIGURE 4.3
Lateral flow assay architecture within a RDT testing cartridge. (Source: US National Aeronautics and Space Administration.)

home-based testing, the specimen is suspended in media and physically manipulated by squeezing nasal swabbed content into the solution.

SARS-CoV-2 RDT relies on paper-based chromatography and capillary forces to carry specimen solution along a series of absorbing pads contained with the test cartridge (see Figure 4.3):

- A sample pad onto which drops of the specimen solution are placed
- A conjugate pad containing antigens and antibodies to SARS-CoV-2 nucleocapsid protein as the preferred target,[5] with a visible probe as the "reporter" read by the person performing the test
- A nitrocellulose membrane where specific antibody-antigen interactions indicating the presence or absence of the targeted antigen, in this case SARS-CoV-2 nucleocapsid, are indicated by a darkening test line and further conjugation indicating the suitability of the test cartridge as the control line

Results are available within 30 minutes and are self-interpreted by the test performer. All RDTs have very high specificities, so a negative test is likely "truly negative"—making one reasonably sure of not being infected with SARS-CoV-2 in the case of a negative RDT. However, RDTs have low to moderate sensitivities, meaning false negatives are possible and should be confirmed with PCR testing, especially for symptomatic persons or those living in areas with high infection rates.

Using Clinical Tests in Making COVID-19 Diagnoses

PCR tests and RDTs are used for screening and confirmation purposes. These tests differ in terms of their targets, whether trained personnel are needed for test performance, *turnaround time (TAT)*, cost, and analytic sensitivity and specificity, as shown in Table 4.3.

During the earliest phases of the pandemic, PCR was the only laboratory test for SARS-CoV-2 authorized by the Food and Drug Administration (FDA) for emergency use; thus, it was used for both screening and diagnostic purposes. Because experimental studies had shown PCR to have superior analytic performance in finding "truly positive" and "truly negative" infected people, it was initially used as a screening tool among healthy people in public health surveillance programs and as a diagnostic tool by healthcare professionals in their assessments of acutely ill people assumed to have COVID-19.

But this degree of precision is costly and time consuming. PCR requires many processing steps by trained lab personnel, producing lengthy TAT in obtaining results. This long TAT is a barrier to the quick decision making needed at the start of an epidemic, when asymptomatic infected people must be placed in quarantine to prevent further transmission of illness, and their close contacts need to know whether they have been infected.

TABLE 4.3

Comparison of PCR and RDT Tests

Metric	PCR	RDT
Assay target	amplified RNA	SARS-CoV-2 nucleocapsid proteins
Trained personnel	yes	no
Turnaround time	hours	15 to 30 minutes
Cost	moderate to high	low
Sensitivity	high	low to moderate
Specificity	high	high

This kind of information is especially vital for frontline healthcare workers responsible for delivering healthcare services to infected patients. The rapid TAT and low cost of RDTs has made them an appealing tool for increasing access to testing.

Looking Back, Looking Ahead

Michael Mina's mission to use convenient and low-cost RDTs to detect SARS-CoV-2 infection within asymptomatic, healthy-appearing people is an example of how screening tests could be used in public health programs during a fast-moving viral pandemic.

For SARS-CoV-2 infections, screening could occur as part of community surveillance programs in which everyone in a captive population (college campus, apartment complex, or extended social connections of an individual) is subjected to SARS-CoV-2 testing using RDTs. Because of their accessibility, RDTs could also be used as part of personal risk assessments when individuals become aware that they could have been exposed to the virus by contact with infected individuals.

As Bayes's theorem indicates, prevalence—or the likelihood of finding an infection in a community—should be kept in mind when results are interpreted on tested individuals, since all clinical tests have varying degrees of sensitivity and specificity. For instance, in a community with a very low SARS-CoV-2 infection rate, a "positive" result in an asymptomatic person might be a "false positive," and one should be cautious in saying "this tested person has this condition."

The convenience of rapid antigen tests is offset by their relatively lower sensitivity compared to "gold standard" PCR tests. To confirm the diagnosis of a suspected case of SARS-CoV-2 infection, especially when someone has signs and symptoms of disease shared with other illnesses, confirmatory diagnostic tests like PCR are needed.

As SARS-CoV-2 evolves, it may be that RDTs become a frontline defense for screening asymptomatic people in high-prevalence communities or those who come into contact with large numbers of people. PCR tests would play a vital role in confirming the results of RDTs as well as providing accurate diagnoses for symptomatic individuals. However testing models for COVID-19 develop, the lessons they provide will be important foundations for responding to the next emerging infectious disease (EID) the world faces.

FUTURE OUTINGS

To understand how the lessons of COVID-19 can be applied to future EIDs, ongoing excursions into the field must explore:

- The feasibility of deploying large-scale testing regimens among asymptomatic populations
- The level of sensitivity and specificity required for a clinical test to effectively screen for potential infection
- Effective use cases and guidelines for screening and diagnostic tests for rapidly spreading EIDs

Student Research Questions

1. When were PCR tests and RDTs approved by the FDA? How were they distributed, and to whom? What kinds of populations had access to which tests?
2. What are the steps to diagnosing COVID-19 when patients seek medical care for symptoms? How are medical history, physical examination, and clinical tests used?
3. How are clinical tests used for diagnosing COVID-19 compared to how they are used to diagnose other common viral infections?
4. In the context of the COVID-19 pandemic, what are the potential consequences of overdiagnosis due to false positives? What are the possible dangers of misdiagnosis due to false negatives?
5. What large-scale testing programs have been employed during the COVID-19 pandemic? How have these programs affected the spread of disease?

NOTES

1. Michael Mina, Roy Parker, Daniel Larremore, "Rethinking Covid-19 Test Sensitivity—A Strategy for Containment," *New England Journal of Medicine*, November 26, 2020, published online September 30, 2020, doi: 10.1056/NEJMp2025631
2. Regev-Yochay G, Kriger O, Mina MJ, Beni S, Rubin C, Mechnik B, Hason S, Biber E, Nadaf B, Kreiss Y, Amit S. (2022). Real World Performance of SARS-CoV-2 Antigen Rapid Diagnostic Tests in Various Clinical Settings. Infect Control & Hospital Epidemiology, 1–20. doi: 10.1017/ice.2022.3

3. "Healthcare professionals" is an encompassing term for physicians, doctors (doctors of allopathic medicine, MDs, or doctors of osteopathic medicine, DOs), advanced practice providers (APPs), or nurse practitioners (NPs), all commonly called "primary care providers" (PCPs). PCPs are "primary" because they are the first to encounter sick patients and are responsible for making the initial diagnosis of illness.

4. In 1993 Mullis received the Nobel Prize for Chemistry for "contributions to the developments of methods within DNA-based chemistry for his invention of the polymerase chain reaction (PCR) method." The Nobel Prize in Chemistry 1993. NobelPrize.org. Nobel Prize Outreach AB 2022. Tuesday, May 24, 2022. https://www.nobelprize.org/prizes/chemistry/1993/summary.

5. Nucleocapsid protein is highly conserved and not subjected to the same degree of mutation pressure as other SARS-CoV-2 structural proteins. Different RDTs may have one or more SARS-CoV-2 protein targets other than nucleocapsid.

5

Therapeutics

The Field: When a new infectious disease emerges, stopping its spread as soon as possible is crucial. But until effective pharmaceutical therapeutics can be developed, the world is faced with a conundrum. Governments must enact policies to contain infection while mitigating the harmful effects of such measures on the economic and social landscape. At the same time, individuals may cast about for drugs or substances they believe can protect them, regardless of the scientific evidence—or lack thereof—for their safety and efficacy. The scientific method enables the development of effective therapeutic agents rooted in pharmacologic principles and an understanding of how viruses function. Despite unique obstacles to developing antiviral medications, the world now has several valid COVID-19 therapeutic agents. Understanding how such agents are developed, and how they can be used in conjunction with other measures, is crucial to responding to future pandemic-level viral outbreaks.

Field Sightings: *in silico*, *in vitro*, *in vivo*, drug repurposing, non-pharmaceutical interventions (NPIs), Susceptible-Infected-Recovered (SIR) forecasting model, "hammer and dance", H1N1 influenza A virus, scientific method, randomized clinical trials (RCTs), bias, randomization, placebo, blinding, peer-reviewed journals, preprint servers, Consolidated Statement of Reporting Trials (CONSORT), therapeutic agent, ADME, pharmacokinetics, pharmacodynamics, first pass metabolism, bioavailability, viral tropism, monoclonal antibodies, RNA-dependent RNA polymerase, proteases, protease inhibitors, cytochrome P4503A4 (CYP3A4), drug-to-drug interactions.

DOI: 10.1201/9781003310525-6

FIELD EXPEDITION: THE STRANGE CASE OF IVERMECTIN

People living near rivers in sub-Saharan African countries avoid blackflies. They know bites from these pesky insects can cause permanent blindness, a condition known as onchocerciasis. Onchocerciasis is one of 13 neglected tropical diseases (NTDs)[1] for which the World Health Organization (WHO) and private humanitarian groups have developed mitigation control strategies. The Onchocerciasis Control Programme (OCP), operated by the WHO between 1974 and 2002,[2] tried to stop river blindness by doing two things: conducting aerial insecticide spraying campaigns to kill blackflies and distributing quantities of a medication called ivermectin.

Ivermectin's discovery, initially used as an antiparasitic agent in animals and crossed over to humans to first treat onchocerciasis, is a fascinating story.[3]

Before ivermectin became a drug, it was an accidentally discovered chemical compound: a macrocyclic lactone first identified in the bacterium *Streptomyces avermitilis* found in Japanese soil samples.[4] This is the origin story for numerous modern medicines. Chemicals found in plants, soil, and bacteria with potential disease-fighting properties are investigated further for development into drugs.

There is no guarantee that a chemical compound found in the natural world will succeed as a drug for human use. Any chemical taken by mouth, rubbed on the skin, or placed directly into the veins must navigate a series of obstacles in the human body before arriving at the cellular or molecular sites where its chemical properties can actuate their healing effects. "A chemical cannot be a drug ... a drug must get in, move about, hang around, and then get out."[5]

After ivermectin was discovered, it was eventually developed into a medication approved by the Food and Drug Administration (FDA) to treat onchocerciasis and other parasitic infections in humans, cattle, sheep, horses, pigs, and dogs. Ivermectin is FDA-approved to be used only in treating parasitic infections; it was never approved as a possible antiviral agent.

In the early days of the COVID-19 pandemic, however, *in silico*[6] (computer-simulated studies) and *in vitro*[7] (laboratory) experiments seemed to indicate that ivermectin interfered with the attachment of SARS-CoV-2 to human cells, suggesting it might be effective as a possible treatment against SARS-CoV-2. Notably, *in vivo* (animal and human subject) experiments—which are critical in determining the actual safety and effectiveness of a drug in combating a particular pathogen—did not support this conclusion. Nevertheless, the widespread use of ivermectin as a treatment for COVID-19 was largely promoted based on computer-simulated and laboratory observations alone.

During the early pre-vaccine phase of the pandemic, when treatment options were limited, using already FDA-approved medications to treat COVID-19 was advanced as an option, a therapeutic strategy called *drug repurposing*. Because of the lab observation that ivermectin interfered with binding between SARS-CoV-2 and human cells, it was promoted as a repurposed COVID-19 treatment, along with azithromycin, chloroquine, and hydroxychloroquine—all FDA-approved drugs for other purposes but not specifically studied in COVID-19 patients or authorized for this use.

Ivermectin's popularity continued even after COVID-19 vaccines and other treatment options became available. This led Merck, ivermectin's US manufacturer, to issue a blunt statement that there was "no scientific basis for a potential therapeutic effect against COVID-19 from pre-clinical studies; no meaningful evidence for clinical activity or clinical efficacy in patients with COVID-19 disease; and a concerning lack of safety data in the majority of studies."[8]

Regardless of Merck's statement and similar pronouncements from government and professional groups[9], ivermectin's use continued unabated. Retail pharmacies saw a massive increased demand from the pre-pandemic baseline average, which was 3,600 prescriptions a week between March 16, 2019, and March 13, 2020. By January 8, 2021, pharmacies were processing 39,000 prescriptions a week, and that figure rose to 88,000 as of August 13, 2021: a 24-fold increase from pre-pandemic levels[10] (see Figure 5.1).

Though some research trials of ivermectin use in COVID-19 patients began to be conducted, many of these studies were later discredited and retracted from publications due to study design weaknesses, misleading conclusions, or a lack of sufficient numbers of experimental subjects.

By February 2021, the lack of strong scientific evidence for ivermectin use led the National Institutes of Health (NIH) to state:

> there is insufficient evidence ... to recommend either for or against the use of ivermectin for the treatment of COVID-19. Results from adequately

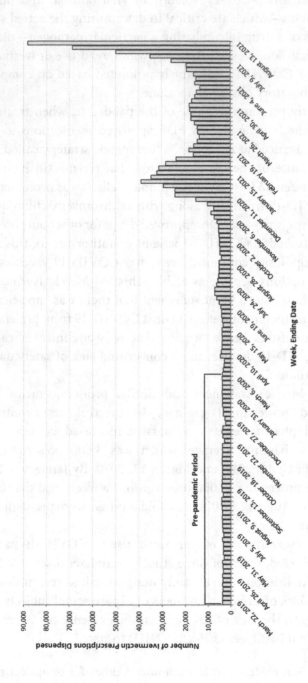

FIGURE 5.1

Estimated number of outpatient ivermectin prescriptions dispensed from retail pharmacies—United States, March 26, 2019–August 13, 2021. (Source: CDC Health Alert Network—CDCHAN 00449.)

powered, well-designed, and well-conducted clinical trials are needed to provide more specific, evidence-based guidance on the role of ivermectin in the treatment of COVID-19.[11]

One year later, results from ivermectin's first "adequately powered, well-designed, and well-conducted clinical trial" was published: a double-blind randomized placebo-controlled trial among 1,358 SARS-CoV-2 patients seen in 12 Brazilian public health clinics. This March 2022 study demonstrated that ivermectin was statistically no different than a placebo in preventing hospitalizations or emergency department visits among SARS-CoV-2 infected patients.[12]

In other words, taking ivermectin had the same effect in treating COVID-19 as a sugar pill.

LESSON LEARNED: THE STRANGE CASE OF IVERMECTIN

Ivermectin is a highly effective therapeutic agent specifically designed to treat parasitic infections in humans and animals. Though computer-simulated and laboratory studies indicated a theoretical ability for ivermectin to interfere with SARS-CoV-2 and human cell receptor binding, these observations were not initially carried forward into the kind of large-scale human clinical trials needed to prove ivermectin should be used to treat COVID-19. At a time when few treatment options were available, ivermectin was culturally promoted as a COVID-19 panacea and continued to be promoted. Even after the kind of large-scale human clinical trial which could produce the strongest evidence of whether a drug should be used to treat an illness proved ivermectin to be ineffective in treating COVID-19—and when effective vaccines and therapies had been developed—demand for the drug only grew.

Fear and misinformation may have driven people to ingest an ineffective drug prior to the development of scientifically sound medications for the treatment of COVID-19, but proven non-drug-based options existed to protect populations—though their true effectiveness was complicated by the economic and social landscape in which the COVID-19 pandemic arose.

CONCEPTS COVERED IN THIS CHAPTER

- Non-pharmaceutical interventions (NPIs)
- The 1918 influenza pandemic
- The scientific method in the development of medicines
- Pharmacologic principles
- Viral life cycle and pathophysiology
- Challenges in developing antiviral medications
- NIH COVID-19 guidelines and treatment recommendations
- NPIs and therapeutics as dual measures

Non-Pharmaceutical Interventions (NPIs)

Prior to the development of effective treatments against this new disease, the only options available were ***non-pharmaceutical interventions (NPIs)***. NPIs in the form of public health measures such as social distancing, facemask wearing, curfews, and quarantine had proven to be effective in containing prior epidemic outbreaks, but enforcing this level of social control raised concerns about the potential for social backlash and the crippling of national economies. China, Italy, and South Korea were the earliest nations to experience the effects of this developing crisis, and each responded differently according to their social, political, and cultural constructs—as would be the case in other nations as the pandemic spread.

While the use of NPIs varied both within and across nations, they allowed health systems more time to better manage hospitalizations, medical supplies, and personnel—and would hopefully provide enough time for vaccines, treatments, and medical supplies to be developed and stockpiled while reducing illness and death in the meantime. But uncertainties remained: which NPIs should be employed, and for how long? When should social distancing be started and stopped? Was there a way to determine this?

At the start of a new epidemic, when real-world data based on actual pathogen transmission, infection cases, and recovery are unclear, mathematical models are designed to forecast possible scenarios that could be used to make public policy decisions. The ***Susceptible—Infected— Recovered (SIR) forecasting model*** is commonly employed to test these scenarios during infectious disease epidemics.

By March 16, 2020, five days after the World Health Organization officially declared SARS-CoV-2 to be a global pandemic, the Imperial College London COVID-19 Response Team[13] developed a simulation model about possible NPIs the United Kingdom and United States governments could employ to prevent their healthcare systems from being completely overwhelmed. They based their model on a 2005 SIR model the team had developed during the H5N1 avian flu pandemic in Southeast Asia. The SARS-CoV-2 model used data from China and Italy, including incubation period, infectiousness profile, basic reproduction rate, hospitalizations, critical care cases, and infection fatality ratios. The model considered a five-month suppression strategy employing the NPIs of home-based isolation, quarantine, social distancing, and school closure. When the simulation was run, however, the results were dire.

The Imperial College Team predicted that, if no mitigation actions were instituted, 510,000 deaths in the United Kingdom and 2.2 million in the United States would occur. Hospitals would be overwhelmed with admissions and critical care bed usage. They proposed that stringent measures would need to be instituted and remain in place until either sufficient herd immunity developed or vaccines were available to prevent transmission from resuming. But according to their simulation model, even with these interventions in place over five months, a case surge would happen months later.

It was hoped, however, that a period of NPIs would allow enough time to develop vaccines that would avoid the projected subsequent surge in infection rates. Learning this, United Kingdom and United States leaders instituted confinement strategies and saw infection cases decline, but lockdowns, quarantines, and social distancing caused devastating consequences for their economies. Public backlash followed—especially in the United States, where cultural values of individual liberty and partisan political forces sowed opposition to the policies—causing leaders to rescind their confinement strategies, inevitably causing infection resurgence. This alternating cycle of confinement and deconfinement was poignantly described as the *"hammer and dance"* by Tomas Pueyo.[14]

NPIs in Another Age: The 1918 Influenza Pandemic

Catastrophe results when human populations are infected with a completely new pathogen. Over 100 years before the SARS-CoV-2 virus arose, the world faced a similar disaster: a completely new H1N1 influenza virus swept around the world beginning in 1918. This pandemic infected nearly 30% of all humans and killed an estimated 50 million, including 675,000 people in the United States—a death toll surpassing the estimated 25 million who died during the Black Death of the fourteenth century.[15]

The first reported cases of the new influenza infection began March 1918 among soldiers at Camp Funston in Fort Riley, Kansas (see Figure 5.2). From Camp Funston, sporadic cases were noted throughout the United States, and the transmissible mutating virus soon spread worldwide in three distinct waves of illness over two years. It killed the largest number of people during its second wave in the fall of 1918, and though cases subsided dramatically by spring 1919, descendants of the virus that caused the pandemic remain in circulation today.

This devastation was caused by a previously unknown *H1N1 influenza A virus*. Among the four known influenza virus types—A, B, C, and D— types A and B cause serious human illness, type C causes mild disease, and type D sickens animals. Influenza A is the only virus which has

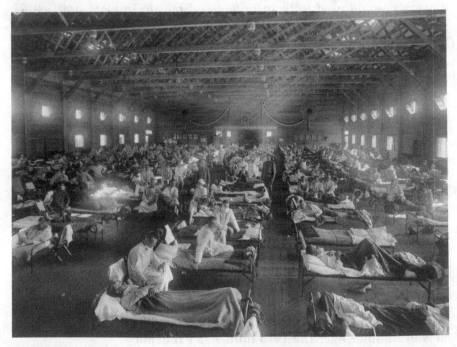

FIGURE 5.2
Soldiers being treated for influenza in 1918 at Camp Funston in Fort Riley, Kansas. (Source: Otis Historical Archives, National Museum of Health and Medicine.)

caused pandemic-level infections. It does so through antigenic changes to its two surface glycoproteins, hemagglutinin (H) and neuraminidase (N). These antigenic protein changes are numerically categorized and called subtypes. H1N1 or H3N2 are the two most common influenza A subtypes circulating today in some form across the southern and northern hemispheres.

Hemagglutinin latches on to human cells to enter the cell. Neuraminidase has three functions: aiding influenza viral entry into cells, helping transport the virus through thick mucus, and facilitating the exit of newly hatched influenza virions from host cells. Today, hemagglutinin and neuraminidase are targets for seasonal influenza vaccines and antiviral medications like oseltamivir (Tamiflu™), which reduce illness severity in all people, especially older persons and those with chronic medical conditions.

But in 1918 there were no vaccines or therapeutics available for use, so NPIs were the only means for limiting the spread of the deadly H1N1 influenza A virus, and they were enacted to various degrees in different regions. American cities that quickly recognized new cases of infection and enacted preventative NPIs by closing schools, entertainments venues, and commercial enterprises reduced rates of death, though this was short-lived, as no city was able to maintain this posture for more than six weeks. Philadelphia saw its first cases in mid-September 1918 but did not institute bans on public gatherings for nearly a month and saw its peak influenza-related deaths rise to 257 in 100,000 people. St. Louis, in contrast, saw its first cases on October 5, 1918, and enacted closures two days later, experiencing a much lower death rate of 31 in 100,000 people (see Figure 5.3).

Further complicating efforts to curtail the spread of disease was obfuscation of data by some nations. The label "Spanish flu" is associated with the 1918 influenza pandemic because Spain, a neutral country not militarily involved in World War I, truthfully reported its ongoing experience with the disease, including the near-fatal illness of its king, Alfonso XIII,[16] at a time when other European nations and the United States were actively suppressing accounts of the spreading illness to maintain military morale during the last months of World War I. President Woodrow Wilson went so far in suppressing disease information as to encourage Congress to enact the short-lived Sedition Act of 1918, curtailing free speech. Ironically, Wilson contracted the influenza strain

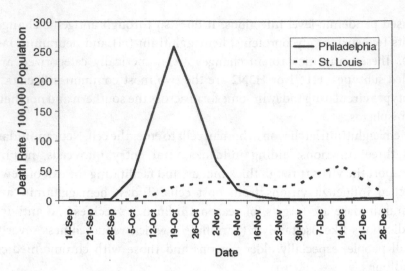

FIGURE 5.3

Peak influenza-related deaths per 100,000 for Philadelphia (solid line) and St. Louis (dashed line) during the second wave of 1918 influenza. (Source: Copyright 2007, National Academy of Sciences, United States.)

during a third wave of infections in the spring of 1919 while in Paris attending peace talks to end the war.

While strains of the virus that caused the 1918 influenza pandemic still circulate today, the development of vaccines and therapeutics have drastically reduced the incidence and virulence of infection.

The Path to Scientifically Sound Therapeutics

Scientific truth is neither relative nor subjective. For hundreds of years, scientific inquiry has been guided by principles intended to establish objective facts, and the methods by which inquiries have been carried out have continually evolved to place more stringent demands on researchers in the service of establishing truth. In the morass of information and misinformation surrounding COVID-19, these principles were crucial in guiding officials, healthcare professionals, and the public to reliable facts about treating the disease.

Scientific Method

Scientific truth is discoverable, fact-based, measurable, and objective. These tenets, established 400 years ago by Francis Bacon, are the basis of the *scientific method* in which natural phenomena is explained by developing a hypothesis about an observed problem, designing experiments to prove or disprove the hypothesis, interpreting observations from experiments, and using these observations to refine the original hypothesis or to develop new hypotheses.

Randomized Clinical Trials (RCTs)

Today research scientists incorporate the scientific method in designing experiments to test whether chemical agents could be helpful as medications for specific illnesses. There are different ways to answer this question, but the experiments which yield the strongest scientific results are *randomized clinical trials (RCT)*.

Finding out whether a therapeutic agent is effective in treating an illness means answering the cause-and-effect question "Does therapeutic agent A result in clinical outcome B?" This hypothesis is the starting point around which experiments are designed. Well-designed RCTs begin with an even more specific cause-and-effect question (e.g., "Does *a specific dose of* therapeutic agent A result in *reducing hospitalizations as* clinical outcome B?").

Well-designed experiments do their utmost to avoid introducing *bias*, which is "any trend or deviation from the truth in data collection, data analysis, interpretation and publication which can cause false conclusions."[17] Bias is reduced when RCTs incorporate two elements:

- *Randomization* to ensure "treatment" and "control" groups are randomly distributed with subjects having similar characteristics (such as age, gender, race), usually by computer algorithm, with the "control" group often receiving a *placebo*, an inert substance similar in appearance to the therapeutic agent being tested.
- *Blinding* whereby all participating parties in the experiment— subjects, investigators, analysts, and monitors—are prevented from knowing who was randomly assigned to which groups.

Peer-Reviewed Journals

After RCTs are conducted, results are published in *peer-reviewed journals* with well-established standards and publication criteria. "Peer-reviewed" means that findings from experiments described in journal articles are scrutinized by others from fields of study in which experiments were conducted. Some peer-reviewed journals have existed for hundreds of years, two of the oldest being the *New England Journal of Medicine*, founded in 1812, and the *British Medical Journal*, founded in 1840.

In a rush to push information about COVID-19 to the public as quickly as possible, results from scientific experiments were often uploaded to **preprint servers**, like the biomedical preprint server bioRxiv maintained by Cold Spring Harbor Laboratory. These articles are not peer-reviewed, though some are submitted to peer-reviewed journals for final publication.

Consolidated Statement of Reporting Trials (CONSORT)

As the volume of conducted RCT studies resulted in thousands of articles being published in hundreds of journals, making determinations about which published studies were worth reading became nearly impossible. By 1996, research scientists and peer-reviewed journal editors agreed upon using the **Consolidated Statement of Reporting Trials (CONSORT)** to describe design elements of RCTs, allowing readers to quickly assess the validity and applicability of clinical trials.

CONSORT uses a flow diagram to describe eligibility criteria for enrolling participants, number of participants randomized to treatment and control groups, number of participants lost to follow-up, and number of participants analyzed. This framework increases transparency and comprehension of RCTs for scientists and the public who otherwise face sifting through a mountain of scientific literature.

Pharmacologic Principles

A disease process ensues when the body experiences damaging change from its normal physiologic state, setting into motion a series of events meant to restore the body back to normal homeostasis. If homeostasis cannot be achieved, the havoc created by the initiating event (e.g., infection by a virus or poor oxygenation from fluid-filled lungs) creates a pathophysiologic state, leaving in its wake damaged cells and tissues unable to function normally.

A **therapeutic agent** (medication, medicine, or drug) is a chemical or biological substance which can prevent, alter, or reverse a disease process. The term "drug" may be associated with negative connotations, but "therapeutic" always implies a restoration to health. Chemical agents may be small molecules like aspirin or large biological proteins like insulin.

Reversing a disease process involves a therapeutic agent "getting in, moving about, hanging around, and then getting out."[18] This is accomplished through a series of tasks known by the acronym *ADME*: absorption (getting in), distribution (moving about), metabolism (hanging around), and excretion (getting out). ADME is the framework for the pharmacology tenet called *pharmacokinetics*. The second pharmacology tenet describing how a therapeutic agent acts to reverse a disease process at the cellular or molecular level is called *pharmacodynamics* (see Figure 5.4).

Medications enter the body by enteral routes (entry through the gastrointestinal tract by mouth) or parenteral routes (entry outside the gastrointestinal tract through muscle or vein injection).

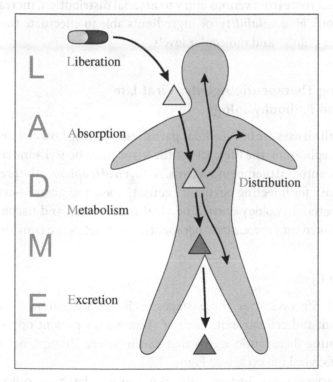

FIGURE 5.4
The pharmacokinetic principles of ADME. (Source: Wikimedia Commons/licensed under CC BY 4.0.)

Though enteral medications taken by mouth are easier to administer than parenteral ones, they need to navigate an obstacle course through the body to reach the arterial circulation for distribution throughout the body: (i) surviving the acidic environment of the stomach to be passed into the small intestine; (ii) entering venous circulation routed to the liver, where *first pass metabolism* occurs, eliminating most of the drug; and (iii) continuing into arterial circulation, where a medication can reach its treatment target.

The liver is primarily responsible for metabolizing medications, which are evacuated through the lower intestine or excreted in urine through the kidneys. Since the liver and kidneys are primarily responsible for metabolizing and excreting medications, any liver or kidney damage will block elimination of drugs, potentially raising the concentration of enteral medications to toxic levels within the body.

Parenteral medications, on the other hand, bypass the gut and have a more direct route from venous entry to arterial distribution, increasing the medication's *bioavailability*, or ingredients able to effectuate therapeutic changes at cellular and molecular levels.

Designing Therapeutic Agents: Viral Life Cycle and Pathophysiology

Because all viruses are intracellular parasites requiring host cell machinery for viral replication, the life cycle of the virus must be well-understood in designing antiviral therapeutic agents. Also, *viral tropism*—the preferences viruses have for infecting particular cells, tissues, and hosts[19]—means the specific pathophysiology created by viral-induced cell and tissue damage must be taken into account as therapeutic approaches are considered.

Viral Life Cycle

The viral life cycle has three stages: cell surface entry, intracellular replication, and cellular exit. Each of these stages present opportunities for designing therapeutic agents that can interfere, disrupt, or terminate the normal viral life cycle (see Figure 5.5).

Cell surface entry begins with the virion[20] latching onto cellular receptors. This attachment can be hindered by immune system components actively created in the body by vaccines or passively administered to the

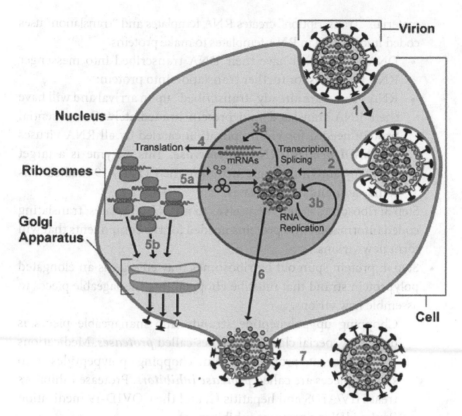

FIGURE 5.5
Virus life cycle showing virion entry, envelope uncoating, release of genetic material, use of intracellular machinery (ribosomes, endoplasmic reticulum, Golgi apparatus) to create new virions, and virion exit. (Source: National Institutes of Health/Public Domain.)

infected person in the form of *monoclonal antibodies* like the COVID-19 monoclonal antibody bebtelovimab.[21]

Intracellular replication of virions occurs in sequential steps. Therapeutic agents interrupting any of these steps will stop viral replication:

- Step 1: inside the cell, the protective envelope which transports the virion is removed, a process called uncoating. Viruses are classified as either DNA or RNA viruses based on their genetic material. SARS-CoV-2 is an RNA virus. Uncoating releases the DNA or RNA genetic material carried by the virus into the intracellular cytoplasm of the host cell.
- Step 2: different "transcription" or "translation" steps are now followed for DNA viruses and RNA viruses after uncoating releases genetic

material. "Transcription" creates RNA templates and "translation" uses coded information in RNA templates to make proteins.

- DNA viruses will have their DNA transcribed into messenger RNA templates for further translation into protein.
- RNA viruses are already "transcribed" upon arrival and will have their RNA translated into protein immediately. One essential protein needed for viral replication carried by all RNA viruses is *RNA-dependent RNA polymerase*. This enzyme is a target for many antiviral drugs, including the COVID-19 medications remdesivir and molnupiravir.
- Step 3: ribosomes fasten themselves to mRNA templates, translating coded information into proteins needed for the components that will form new virions.
- Step 4: protein spun out by ribosomes may emerge as an elongated polyprotein strand that must be chopped into manageable pieces to assemble new virions.
  - Chopping up polypeptide strands into manageable pieces is done by a special class of enzymes called *proteases*. Medications which prevent proteases from chopping polypeptides into smaller pieces are called *protease inhibitors*. Protease inhibitors treat HIV/AIDS and hepatitis C, and the COVID-19 medication Paxlovid™ is a protease inhibitor.
- Step 5: chopped-up polypeptides are further processed by host cell organelles, the endoplasmic reticulum and Golgi apparatus, and packaged into new virions, a process called assembly.
- Step 6: newly assembled virions are now ready to exit the host cell. Exiting presents another opportunity for antiviral therapeutic action.
  - This is the mechanism of action for all neuraminidase inhibitors like Tamiflu® (oseltamivir) used to treat influenza infection. Neuraminidase disentangles newly formed influenza virion particles from host cells. In the presence of neuraminidase inhibitors like oseltamivir, these virion particles remain tethered to host cells and cannot be released into circulation to do further damage.

Viral-Induced Pathophysiology

Viral infections damage different cells and tissues throughout the body, and various disorders in normal bodily function occur depending on where this happens. For instance, many viruses are transmitted from person to person by breathing in the exhaled air from an infected person. These respiratory transmitted viruses may primarily infect the upper respiratory tract, causing cough, or travel more deeply into the lower respiratory tract, causing pneumonia in the lungs.

SARS-CoV-2 uses the ACE2 receptor to enter human cells. ACE2 receptors are located throughout the body and are especially prominent in the respiratory system. This entry point into the upper respiratory system explains the loss of smell and taste and dry cough that are commonly experienced as the first symptoms of COVID-19. But SARS-CoV-2 may travel deeper into the lungs, causing pneumonia severe enough to interfere with oxygenation.

However, COVID-19 is not just a respiratory illness; it is a multisystem disorder affecting many organs. Seriously ill COVID-19 patients often develop blood clots and thromboembolism, which is why anticoagulation agents like heparin are given to COVID-19 hospitalized patients.

Viral infections trigger remarkable immune responses intended to beneficially rid the body of the invading virus. But when an overactive immune system is not turned off, or "downregulated," it continues to bathe internal organs in damaging immunoproteins and enzymes.

The body naturally produces downregulating substances called corticosteroids meant to reduce overactive inflammatory responses. Corticosteroid compounds have been chemically created as therapeutic agents and have many medical uses. Dexamethasone is one such corticosteroid with potent anti-inflammatory activity and is now used to treat seriously ill COVID-19 patients.

Three Challenges in Developing Antiviral Medications

An unfathomably large number of some 10^{31} viruses exist on Earth,[22] constituting an invisible virosphere on our planet. But only a minuscule fraction of these—some 300 viruses, mostly from animal sources—are believed to have potential for causing human disease.[23]

Among hundreds of potential viral pathogens, antiviral drugs have only been developed against nine viruses: human immunodeficiency virus (HIV), hepatitis B virus (HBV), hepatitis C virus (HCV), herpesvirus, influenza virus, human cytomegalovirus (CMV), varicella-zoster virus (VZV), respiratory syncytial virus (RSV), and human papillomavirus (HPV).[24]

This paucity of antiviral medications may exist for three reasons:

- Viral biology: viruses are intracellular parasites, essentially hiding within host cells. This makes them notoriously difficult antimicrobial targets compared to bacteria, fungi, and other microorganisms, whose cell structures and metabolism distinguish them from human cells, making targeted antimicrobial development easier.
- Pharmacoeconomics: billions must be spent on research and development (R&D) to bring a new therapeutic agent to market.[25] Pharmaceutical companies generally cannot realize the same return on investment for antiviral drugs as they can for other therapeutic classes. This economic reality may explain why oncologic, diabetes, and autoimmune therapeutic classes ranked first, second, and third for R&D spending in 2019, at nearly $70 billion for each drug class, compared to R&D spending of about $24 billion for HIV antivirals and $6 billion for viral hepatitis drug classes.[26]
- Biosafety laboratory infrastructure: as the number of pathogenic emerging viruses increases, there is a greater need for physical infrastructure and trained personnel to safely handle and study these dangerous pathogens within biosafety laboratories (BSLs). Construction and human capital costs for developing BSLs are significant and may impede viral research, further hampering antiviral medication development.

At the onset of the COVID-19 pandemic, scientists leaped into action on multiple fronts, including the development of antiviral medications. With a never-before-seen virus, researchers must first build the groundwork for the development of therapeutic agents: understanding the viral life cycle and pathophysiology of the virus. As these features of SARS-CoV-2 were identified and further studied, therapeutic development based on the scientific method proceeded. Despite the challenges endemic to creating antiviral medications, a number of viable options emerged; as ongoing research revealed more about the virus and the effectiveness of various therapeutics, new medications were produced. The NIH became a crucial source of reliable information and guidance for healthcare workers and the public.

National Institutes of Health COVID-19 Treatment Guidelines

The COVID-19 infodemic and the COVID-19 pandemic began in lockstep. Barely two months into the new infectious disease outbreak, healthcare providers were inundated with confusing and conflicting advice about drug repurposing to develop treatment protocols for COVID-19 patients.

To squelch this confusion, the Secretary of Health and Human Services and the White House Coronavirus Task Force asked the NIH to convene an advisory panel comprised of experts from academia, professional societies, and the federal government to review all available evidence about specific therapeutic agents and render recommendations about their use. The group was quickly assembled, began its work, and by April 21, 2020, had posted its first set of "NIH COVID-19 Treatment Guidelines."

The guidelines rate therapeutic agents with an A, B, or C letter grade based on the strength of the panel's rendered recommendation or a I, IIa, IIb, or III Roman numeral grade based on the quality of available evidence at the time the panel reviewed the therapeutic agent.

The panel ranks these rated assessments in four categories:

- "Recommends for the treatment of COVID-19 (rating)" describes therapeutic agents with evidence of efficacy and safety based on large cohort studies or randomized clinical trials.
- "Recommends against the treatment of COVID-19 (rating)" is applied in cases where data indicate safety concerns or no treatment benefit.
- "Insufficient evidence" (no rating provided) is applied to therapeutic agents for which no evidence is available to rate.
- "Recommends against use except in a clinical trial" leaves the way open for investigational use of a therapeutic agent that is not yet recommended for use.

The guidelines have been updated numerous times since their initial posting to keep pace with drug development and clinical trial results. A recent NIH study showed that the guidelines have attained 17 million page views since inception, with many international views.[27] Amidst the deluge of information and misinformation about COVID-19 therapeutic options, these guidelines have provided healthcare professionals much-needed clarity as research evolves and new therapeutics are developed.

NIH COVID-19 Treatment Recommendations

Patients with SARS-CoV-2 now have a number of therapeutic options available to them based on a better understanding of the life cycle of the virus and the pathophysiology it creates after infection. The goal of using COVID-19 medications is to prevent serious disease progression and death.

The NIH COVID-19 Treatment Guidelines Panel continues to make recommendations about tailored therapeutic approaches to COVID-19 management based on reviewing the available scientific evidence. When considering therapeutic approaches, the following should be kept in mind:

- The strongest sources of scientific evidence about whether particular drugs should be used for treatment are randomized clinical trials or large observational studies.
- Different therapeutic interventions will be needed for non-hospitalized and hospitalized patients.
- An equitable way to allocate medications based on disease progression risk (immunocompromising or chronic medical conditions) and vulnerability (age or unvaccinated status) may be needed if supplies of therapeutic agents are scarce.
- Logistical challenges (such as giving IV medications to non-hospitalized patients over many days) may complicate therapeutic approaches.
- As SARS-CoV-2 mutates to more transmissible forms, therapeutic agent effectiveness will need to be reassessed as shown by resistance to certain monoclonal antibodies after their initial use.

The available treatment options recommended by the NIH Panel for non-hospitalized SARS-CoV-2 infected people are compared in Table 5.1. These options are ranked according to risk reduction (progression to severe COVID-19, hospitalization avoidance, or death) and the clinical trial in which this was demonstrated.

Paxlovid™ and remdesivir are recommended as preferred therapies by the NIH. Bebtelovimab and molnupiravir are recommended as alternative therapies.

Paxlovid™ (Ritonavir-Boosted Nirmatrelvir)

Paxlovid™ is a protease inhibitor drug cocktail containing ritonavir and nirmatrelvir. Ritonavir is a "boosting agent" added to increase blood levels of the active protease inhibitor nirmatrelvir.

TABLE 5.1

COVID-19 Therapeutic Options for Ambulatory Patients

Medication	Risk reduction	Clinical trial	Dose	Caveats
Paxlovid™	89%	EPIC-HR	two 150 mg nirmatrelvir tablet one 100 mg ritonavir tablet twice daily for five days	drug-to-drug interactions
Remdesivir	87%	PINETREE	200 mg IV day 1 100 mg IV days two and three	monitored setting
Bebtelovimab	not known	Phase 2 RCT	175 mg IV single infusion	monitored setting
Molnupiravir	30%	MOVe-OUT	800 mg tablet twice daily for five days	mutagenic risk

(*Source:* NIH COVID-19 Treatment Guidelines, April 8, 2022.)

Protease inhibitors are a well-known therapeutic class of medications used predominantly to treat HIV/AIDS. From years of HIV/AIDS experience, where scarcity of effective therapies meant paying close attention to drug resistance, it was known that "boosting" protease inhibitors with a companion drug able to dampen liver metabolism maintains steady protease inhibitor levels. Maintaining steady drug levels prevents viral mutations, as fluctuating drug levels allow viruses to find new defenses against medications.

Ritonavir is the boosting agent used most often in protease inhibitor drug cocktails. It inhibits *cytochrome P4503A4 (CYP3A4)*, the liver enzyme which metabolizes most medications. But the enzyme decrement caused by ritonavir affects the metabolism of all other therapeutic agents and will increase their blood levels possibly causing toxicity. Because of possible toxicity, prescribers of protease inhibitor drug cocktails must take careful histories of all medications, over-the-counter agents, or nutritional supplements, altering their doses to avoid dangerous *drug-to-drug interactions.*

Nirmatrelvir stops an essential step in viral replication. After uncoating, SARS-CoV-2 releases its single strand of RNA which is immediately translated into two polyproteins pp1a and pp1ab. These polyproteins contain all individual components to make new SARS-CoV-2 virions and must be chopped into smaller pieces to be activated. The chopping function is performed by M^pro, the main protease of SARS-CoV-2. Nirmatrelvir blocks the actions of this protease.

The EPIC-HR[28] trial (Evaluation of Protease Inhibition for Covid-19 in High-Risk Patients) found an 89% reduction in COVID-19 related hospitalization or death in the nirmatrelvir-treated group when compared to placebo among 2,246 randomized trial participants.[29]

This remarkable outcome combined with historical experience with protease inhibitors explains why Paxlovid™ is the leading preferred antiviral treatment for non-hospitalized COVID-19 patients, keeping in mind its significant drug-to-drug interactions from CYP3A4 inhibition by ritonavir which dampens liver metabolism of all medicines.

Remdesivir

Remdesivir remains the only fully FDA-approved medication specific for treating COVID-19 in hospitalized and non-hospitalized adult and pediatric patients. All other COVID-19 medications have emergency use authorization only and continue to be studied, producing results that will eventually be submitted to FDA for new drug license approval as they continue to be given as emergency-authorized therapies to patients.

Remdesivir has its origins in the 2014 West Africa Ebola epidemic, the largest Ebola outbreak, causing 11,325 deaths among 28,652 people mostly in Sierra Leone (14,124 cases), Liberia (10,678 cases), and Guinea (3,814 cases).[30] Remdesivir was found to not only be effective against Ebola, an RNA virus, but other RNA viruses including coronaviruses.[31]

RNA viruses need to transcribe their genetic material into RNA templates to start replication. This requirement to transcribe viral RNA genetic material into RNA templates useful for replication is why all RNA viruses code for the RNA-dependent RNA polymerase (RdRp) protein. Remdesivir is a counterfeit nucleotide analogue which RdRp treats as an authentic nucleotide as it performs its transcribing function. When phony remdesivir is incorporated by RdRp as if it were a "real nucleotide" it stalls RdRp function, bringing viral replication to a halt.

The PINETREE clinical trial found an 87% reduction in COVID-19 related hospitalization or death in the remdesivir-treated group when compared to placebo among 562 randomized trial participants.[32] Remdesivir is administered intravenously to non-hospitalized patients on three consecutive days, and patients must be encouraged to complete all three days of treatment. Though comparable in risk reduction, the inconvenience of outpatient IV administration likely explains why remdesivir is second to Paxlovid™ as the preferred COVID-19 therapy.

Bebtelovimab

Bebtelovimab is a monoclonal antibody (MAB), and was the only monoclonal antibody (MAB) recommended for use by the NIH COVID-19 Treatment Panel as of April 2022. Use of monoclonal antibodies has been challenged by mutating SARS-CoV-2 variants, which differ from region to region. In an April 29, 2022 publication, only Phase 2 clinical trial data was available for bebtelovimab, but its mechanism of action is similar to other monoclonal antibodies studied in Phase 3 trials, and this became the basis for its recommended use.

COVID-19 MABs are designed to target a particular configuration of spike glycoprotein on cell surfaces of SARS-CoV-2. Most SARS-CoV-2 variants acquire numerous mutations in the genomic sequence coding for spike glycoprotein, which changes its configuration and helps the mutating variant evade neutralization by monoclonal antibodies targeted against it.

Other MABs previously recommended for use (bamlanivimab plus etesevimab, casirivimab plus imdevimab, and sotrovimab) are no longer recommended due to their loss of effectiveness against the most current circulating SARS-CoV-2 variants. Because of the increasingly changing nature of SARS-CoV-2 variants, MABs will likely continue as an alternative rather than preferred therapy.

Molnupiravir

Molnupiravir is another nucleotide analogue that affects normal transcription carried out by RNA-dependent RNA polymerases, but its actions induce mutations which could possibly occur in human cells. The MOVe-OUT clinical trial found a 30% reduction in COVID-19 related hospitalization or deaths in the molnupiravir-treated group when compared to placebo among 1,443 randomized trial participants.[33] Its 30% risk reduction and mutagenic possibilities make molnupiravir an alternative therapy, but if other treatments are not available for persons likely to experience serious outcomes or death from COVID-19, molnupiravir should be considered an option.

NPIs and Therapeutics as Dual Measures

In the absence of pharmaceutical agents to prevent and control the spread of disease, NPIs are the only option. As evidenced by the massive amount of death caused by the 1918 influenza pandemic, the effectiveness of NPIs is contingent on the ability of governments and societies to accept and implement such measures. This was on display during the early days of the COVID-19 pandemic—the hammer and dance that led to fluctuating levels of infection.

However, even with the availability of vaccines and therapeutics, some NPI measures will likely be needed due to virus mutations. In order for NPIs and pharmaceutical interventions to be effectively used in conjunction, a balance must be struck. Extreme enactment of NPIs can backfire, as was the case with China's "Zero-COVID" policy of spring 2022. This policy was enacted due to a March 2022 rise in cases in Shanghai, which led to a shutdown that continued for two months, not only restricting movement of its citizens but worsening global supply-chain challenges, as Shangha-based manufacturers were not able to operate normally.[34] In a May 10, 2022 briefing, the WHO Secretary-General said that China's policy was "not sustainable considering the behavior of the virus now and what we anticipate in the future."

Looking Back, Looking Ahead

Early in the COVID-19 pandemic, comparisons to the 1918 influenza pandemic were common in the media, and for good reason: both circumstances involved a novel virus that spread rampantly and for which there was no effective treatment. And in both cases, governments had to decide on measures that could protect populations from multiple types of harm: sickness and death, economic turmoil, the erosion of social institutions. Individuals, too, made choices about how they would protect themselves in an uncertain landscape—often relying on emotional appeals rather than scientific evidence.

With the advent of promising therapeutics, the equation has changed. Those who contract COVID-19 have pharmaceutical treatment options that may reduce their chances of severe sickness and death; the promise of more therapeutics seems to be on the horizon. When people believe that infection is less dangerous, how does their behavior change? How do governments shift the balance of protecting public health with NPIs while supporting economic health with relaxed restrictions? What trade-offs are people willing to accept?

The development of therapeutics may have curbed some of the fear in the public consciousness, but the landscape of COVID-19 remains uncertain. The mutating virus has become more transmissible, and the effectiveness of therapeutics on future variants is unknown. The challenges to developing viral therapeutics are unlikely to be reduced. When the next emerging infectious disease (EID) arises, what role will NPIs and therapeutics play in the world's response?

FUTURE OUTINGS

To understand how the lessons of COVID-19 can be applied to future EIDs, ongoing excursions into the field must explore:

- The prevalence and effects of drug repurposing in response to infectious agents for which vaccines and pharmaceutical therapeutics do not exist
- The historical effects of various NPIs on infection containment, economic activity, and social functioning
- The resources and strategies necessary to overcoming challenges to developing antiviral medications

Student Research Questions

1. What scientific evidence was cited to support the use of ivermectin as a therapeutic agent for COVID-19? What made the arguments for its use so appealing to people? Why did countervailing evidence not curb its use?
2. What other agents were advanced for the treatment of COVID-19 in the early days of the pandemic? Which had valid scientific backing and which did not? How did the actions of public figures, media outlets, and social media affect the use of these agents?

3. What factors must be considered to determine if drug repurposing is scientifically valid? When and how do healthcare providers prescribe FDA-approved drugs for off-label use?

4. What NPIs were used to curtail the spread of COVID-19 prior to the development of therapeutics? How effective were they in different countries and regions, and what factors contributed to those effects? How does this compare to NPIs used in earlier pandemics?

5. How do antiviral therapeutics compare to antibiotics in terms of development, function, and effectiveness?

6. What therapeutics exist for the treatment of viruses other than SARS-CoV-2? How and when were they developed? What has been the effect on the infection rates of the diseases they treat?

NOTES

1. "Neglected" or disregarded illnesses afflicting some one billion people in poorer nations, most living in tropical climates; "neglected tropical diseases" are often spread by insects (vector-borne transmission) in communities with poor sanitation and unsafe drinking water. Other NTDs include ascariasis, trichuriasis, hookworm infection, schistosomiasis, lymphatic filariasis, trachoma, leishmaniasis, Chagas' disease, leprosy, human African trypanosomiasis, dracunculiasis, and buruli ulcer. See Hotez, PJ et al, "Control of neglected tropical diseases," *New England Journal of Medicine*, September 6, 2007, volume 357, number 10, pages 1018–1027.

2. Onchocerciasis Fact Sheet, WHO, June 14, 2019, accessed October 5, 2021, www .who.int/news-room/fact-sheets/detail/onchocerciasis.

3. Crump A, Omura S. "Ivermectin, 'wonder drug' from Japan: the human use perspective." *Proc Jap Acad Ser B Phys Biol Sci.*, 2011; volume 87, number 2, pages 13–28. https://doi.org/10.2183/pjab.87.13, accessed April 28, 2022.

4. National Center for Biotechnology Information. PubChem Compound Summary for CID 6321424, Ivermectin. https://pubchem.ncbi.nlm.nih.gov/compound/ Ivermectin. Accessed April 28, 2022.

5. Hodgson J. "ADMET—turning chemicals into drugs," *Nature Biotechnology*, August 2001, volume 19, pages 722–726.

6. Lehrer S, Rheinstein PH. "Ivermectin docks to the SARS-CoV-2 spike receptor-binding domain attached to ACE2," *In Vivo*, 2020, volume 34, pages 3023–3026, doi:10.21873/invivo.12134.

7. Caly L, Druce JD, Catton MG, Jans DA, Wagstaff KM. "The FDA-approved drug ivermectin inhibits the replication of SARS-CoV-2 *in vitro*," *Antiviral Research*, April 3, 2020, https://doi.org/10.1016/j.antiviral.2020.104787, accessed April 28, 2022.

8. Merck Statement on Ivermectin use during the COVID-19 pandemic, February 4, 2021, www.merck.com/news/merck-statement-on-ivermectin-use-during-the -covid-19-pandemic, accessed October 5, 2021.

9. "AMA, APhA, ASHP statement on ending use of ivermectin to treat COVID-19," AMA Press Release, September 1, 2021. "Ivermectin products are not approved by FDA to prevent or treat COVID-19," CDC Clinician Outreach and Communication Activity, February 1, 2022, https://emergency.cdc.gov/newsletters/coca/020122 .htm, accessed February 17, 2022. "WHO advises that ivermectin only be used to treat COVID-19 within clinical trials," WHO Press Release, March 31, 2021, https:// www.who.int/news-room/feature-stories/detail/who-advises-that-ivermectin-only -be-used-to-treat-covid-19-within-clinical-trials, accessed March 4, 2022.

10. "Rapid increase in ivermectin prescriptions and reports of severe illness associated with use of products containing ivermectin to prevent or treat COVID-19," CDC Health Advisory, August 26, 2021, https://emergency.cdc.gov/han/2021/han00449 .asp, accessed September 2, 2021.

11. COVID-19 Treatment Guidelines Panel. Coronavirus Disease 2019 (COVID-19) Treatment Guidelines. National Institutes of Health. Available at https://www.cov id19treatmentguidelines.nih.gov/. Accessed April 25, 2022.

12. Reis, G, et al. "Effect of early treatment with ivermectin among patients with Covid-19," *New England Journal of Medicine*, March 30, 2022, doi:10.1056/ NEJMoa2115869, accessed April 16, 2022.

13. Neil M Ferguson, Daniel Laydon, Gemma Nedjati-Gilani et al. "Impact of non-pharmaceutical interventions (NPIs) to reduce COVID-19 mortality and healthcare demand," Imperial College London, March 16, 2020, doi:https://doi. org/10.25561/77482.

14. "Coronavirus: The Hammer and the Dance—What The Next 18 Months Can Look Like, if Leaders Buy Us Time," March 19, 2020, https://tomaspueyo.medium.com/ coronavirus-the-hammer-and-the-dance-be9337092b56.

15. Limited historical morbidity and mortality data have led researchers to estimate total number of deaths for the 1918 influenza pandemic from 17.4 million to 100 million, though most accounts use 50 million as a total death estimate. Similar data limita- tions existed in calculating US death totals with 675,000 as total estimated deaths. World population in 1918 was thought to be 1.8 billion and with some 500 million total global infections, 28% of all humans were infected with 1918 influenza.

16. Alfonso likely survived his deadly bout with 1918 influenza from having contracted the 1889–1890 "Asiatic" or "Russian" influenza pandemic when he was three years old and retaining some immunity to the 1918 H1N1 virus.

17. Simundic, AM, "Bias in research," *Biochemia Medica*, 2013, volume 23, number 1, pages 12–15, http://dx.doi.org/10.11613/BM.2013.003, accessed April 27, 2022.

18. See Note 5.

19. McFadden G, Mohamed MR, Rahman MM, Bartee E. "Cytokine determinants of viral tropism," *Nature Reviews Immunology*, September 2009, volume 9, pages 645–655.

20. Whereas "virus" describes various states of obligate intercellular parasites, "virion" is a specific noun meaning the virus outside the host cell existing in extracellular space. As viral replication occurs within infected host cells, new virions are created which will soon be extruded from the cell into extracellular space.

21. Monoclonal antibodies used to treat viral infections all have an ending suffix "-vimab" meaning "viral monoclonal antibody." Prefixes are proprietary names chosen by manufacturers.

22. Rodrigues RAL, Andrade ACSP, Boratto PVM, Trindade GS, Kroon EG and Abrahão JS (2017). "An Anthropocentric View of the Virosphere-Host Relationship," *Front. Microbiol.*, volume 8, article 1673, pages 1–11. doi: 10.3389/fmicb.2017.01673, accessed April 29, 2022.

23. Mollentze N, Babayan SA, Streicker DG (2021). "Identifying and prioritizing potential human-infecting viruses from their genome sequences," *PLoS Biol*, volume 19, number 9, pages 1–25. https://doi.org/10.1371/journal.pbio.3001390, accessed April 29, 2022.

24. De Clercq E, Li G. "Approved antiviral drugs over the past 50 years," *Clin Microbiol. Rev.*, 2016, volume 29, pages 695–747, doi:10.1128/CMR00102-15, accessed July 5, 2021.

25. Wouters OJ, McKee M, Luyten J. "Estimated research and development investment needed to bring a new medicine to market 2009–2018," *JAMA*, 2020, volume 323, number 9, pages 844–853. doi:10.1001/jama.2020.1166.

26. Research and Development in the Pharmaceutical Industry, Congressional Budget Office, April 2021.

27. See Note 11.

28. Clinical trials are often given pithy memorable acronyms as a branding device as in "the EPIC-HR trial."

29. Hammond J et al. Oral nirmatrelvir for high-risk, nonhospitalized adults with COVID-19," *NEJM*, volume 386, number 15, pages 1397–1408, April 14, 2022.

30. CDC, "2014–2016 Ebola Outbreak in West Africa," https://www.cdc.gov/vhf/ebola/history/2014-2016-outbreak/index.html. Accessed May 1, 2022.

31. Eastman RT et al. "Remdesivir: a review of its discovery and development leading to emergency use authorization for treatment of COVID-19," *ACS Cent. Sci.*, 2020, volume 6, pages 672–683, https://dx.doi.org/10.1021/acscentsci.0c00489, accessed May 1, 2022.

32. Gottlieb RL et al. "Early remdesivir to prevent progression to severe Covid-19 in outpatients," *NEJM*, volume 386, number 4, pages 305–315, January 27, 2022.

33. Bernal AJ. "Molnupiravir for oral treatment of Covid-19 in nonhospitalized patient," *NEJM*, volume 386, number 6, pages 509–520, February 10, 2022.

34. In May 2022, GE Healthcare had its iodine radiology contrast manufacturing facility in Shanghai shuttered, stifling supply of this imaging additive and limiting performance of radiologic procedures like angiography.

6

Regulatory Agencies

The Field: Regulatory agencies play a vital role in public health and have been pivotal to the United States' response to the COVID-19 pandemic. While operating within the executive branch of the government, they enjoy broad authority to issue guidance and authorizations in service of protecting the common good. During the pandemic, two agencies in particular—the Food and Drug Administration (FDA) and the Centers for Disease Control and Prevention (CDC)—have been instrumental in shaping public policy. However, lack of coordination among the agencies and the executive branch have also produced public confusion and discord among policymakers.

Field Sightings: White House COVID-19 Response Team, Food and Drug Administration (FDA), Centers for Disease Control and Prevention (CDC), Vaccines and Related Biological Products Advisory Committee (VRBPAC), Advisory Committee on Immunization Practices (ACIP), constitutional federal republic, Interstate Commerce Commission (ICC), Administrative Procedures Act (APA), "notice and comment" process, rulemaking, "fourth branch" of government, Department of Health and Human Services (DHHS), Federal Food, Drug and Cosmetic Act of 1938 (FFDCA), Public Health Service Act of 1944 (PHSA), emergency use authorization (EUA), medical countermeasures (MCM).

FIELD EXPEDITION: THE BOOSTER ANNOUNCEMENT

Upon his inauguration as the 46th president of the United States in January, 2021, President Joe Biden established the *White House COVID-19*

DOI: 10.1201/9781003310525-7

Response Team, whose members included Dr. Anthony Fauci, Chief Medical Adviser to President Joe Biden; Dr. Rochelle Walensky, Director of the Centers for Disease Control and Prevention (CDC); Vice Admiral Dr. Vivek Murthy, the Surgeon General; and Jeffrey Zients, coordinator of the Biden administration's COVID-19 response.

On Wednesday, August 18, 2021, the Response Team held a press conference during which Murthy stated:

> today, we are announcing our plan to stay ahead of this virus by being pre-pared to offer COVID-19 booster shots to fully vaccinated adults 18 years and older. They would be eligible for their booster shot eight months after receiving their second dose of the Pfizer or Moderna mRNA vaccines. We plan to start this program the week of September 20th, 2021.[1]

This announcement seemed to be a fully formed operational plan identifying a booster vaccination group (adults 18 years and older completing a two-dose series of the Pfizer or Moderna vaccine), timing for when the booster was to be given (eight months after a second Pfizer or Moderna vaccine), and a starting date for boosters (September 20, 2021).

The timing of the announcement was sudden, especially coming from a new presidential administration promising to "follow the science" in matters related to the pandemic.[2] "Following the science" meant adhering to a well-known sequence of steps, including regulatory reviews by the *Food and Drug Administration (FDA)* and *Centers for Disease Control and Prevention (CDC)*, the federal agencies responsible for overseeing vaccine, medication, and medical device use by the American public.

If expanded use of previously authorized COVID-19 vaccines as "booster shots" was to occur, the first step in making this determination would be having Pfizer or Moderna submit results from booster shot clinical trials to the FDA's *Vaccines and Related Biological Products Advisory Committee (VRBPAC)*, an 18-member group of government and private sector experts convened to review the scientific merit of clinical trial data, developing recommendations, holding a committee vote, and sending its recommendations and vote results to the FDA Commissioner for final approval.

The CDC would follow the same procedure as the FDA, but instead of reviewing the scientific merits of clinical trial data as performed by the FDA, the CDC's *Advisory Committee on Immunization Practices (ACIP)*, a 15-member group of government and private sector experts, would review the clinical and logistical aspects of booster administration, develop recommendations, vote, and send these recommendations and vote results to the CDC Director for final approval.

These sequenced steps were well known to the public by the time of the August 2021 booster announcement. However, none of the established steps were followed by the time of the White House COVID-19 Response Team's announcement.

On September 17, 2021, one month after the booster announcement, the FDA's VRBPAC held a meeting to review Pfizer's booster data for individuals 16 years of age and older. It did not recommend boosters for all adults 18 years and older, as the White House announcement had stipulated. Instead, it recommended booster doses for only two groups: those 65 and older and anyone at high risk of occupational exposure to COVID-19.

A week later, during a September 23 meeting, the CDC's ACIP considered the same data regarding Pfizer booster shots. The ACIP not only disagreed with the White House blanket plan to boost all adults 18 years and older, but it also disagreed with VRBPAC's recommendation of boosters for those at high risk due to occupational exposure.

The ACIP voted to recommend booster doses for only three groups, and its vote tallies indicated some discord among its 15 voting members: members voted unanimously for boosting those 65 years of age and older and anyone living in skilled nursing facilities. However, vote tallies for boosting two other groups having certain medical conditions reflected disagreement: boosting adults 50 to 64 years old passed by a 13 to 2 vote tally, and boosting adults 18 to 49 years old passed by a 9 to 6 vote tally.

Adding to the confusion, the CDC Director disagreed with the ACIP in its recommendation to exclude occupational workers from receiving boosters and only boosting three other groups. Traditionally, CDC Directors accept all ACIP agreed-upon recommendations, but Walensky's press release about the ACIP vote stated:

> I believe we can best serve the nation's public health needs by providing booster doses for the elderly, those in long-term care facilities, people with underlying medical conditions, and for adults at high risk of disease from occupational and institutional exposures to COVID-19. This aligns with the FDA's booster authorization and makes these groups eligible for a booster shot.[3]

Walensky's public disagreement with her only advisory committee was an attempt to harmonize opinion between FDA and CDC, but even this was contrary to the stated goals of the booster announcement. Some suggested the precipitate timing of the White House COVID-19 Response Team's August 18 statement, issued before any regulatory reviews had occurred, was an attempt to deflect media attention away from the poorly conducted troop withdrawal from Afghanistan, which ended the 20-year long war in Afghanistan on August 30, 2021.

LESSON LEARNED: THE BOOSTER ANNOUNCEMENT

The announcement by the White House Response Team to expand use of COVID-19 vaccines to include booster doses to all adults 18 years and older leapfrogged the usual protocol in how these decisions were typically made. All members of the Response Team knew advisory committees to FDA and CDC—comprised of scientific, clinical, and policymaking experts— would need to carefully review clinical trial data from any vaccine manufacturer seeking changes to how its product would be used by the American public. Instead of awaiting these deliberations, the Response Team prematurely announced a booster plan that was subsequently contradicted by recommendations from FDA and CDC advisory groups. Attempts to rectify these missteps only increased public confusion and mistrust.

While regulatory agencies are part of the executive branch of the federal government, they are not beholden to the agenda of a given presidential administration. Agencies are established and imbued with certain powers by acts of Congress. The heads of these agencies are appointed by the president and may, therefore, change every four to eight years, but they are staffed by career bureaucrats whose work is focused on a much longer view. Inevitably, tensions may arise between the political concerns of a given White House and the priorities of agencies. This is the messy process of a constitutional federal republic: we are governed by the rule of law, with checks and balances in place to mediate the conflicts among different aspects of the government. In the face of a public health crisis, these conflicts are often laid bare—an issue that must be grappled with as crises arise in the future.

CONCEPTS COVERED IN THIS CHAPTER

- The structure of the United States government and the role of regulatory agencies
- The origins and legal framework for regulatory agencies
- The Centers for Disease Control and Prevention (CDC)
- The Food and Drug Administration (FDA)
- Emergency use authorization (EUA)

The United States: A Constitutional Federal Republic

The United States is a *constitutional federal republic*, which means the Constitution is the supreme law of the land for a nation with a centralized federal government performing essential functions on behalf of a republic ruled by the will of the people.

As supreme law of the land, the Constitution describes how the United States is governed by three branches of government—executive, legislative, and judicial—each exercising prescribed constitutional authority and interacting with other branches through checks and balances. In essence, the Constitution is a power sharing document describing how governing power is shared and controlled.

Through nearly 250 years of national expansion, population growth and displacement, one civil war, two world wars, and economic depression, the stability of the United States has been tried, and its cohesion has held. The democratic experiment launched in 1776, when 13 colonies declared their independence from British rule, continues today due to principles within the Constitution and the common agreement of its citizens to adhere to the rule of law.

The Constitution establishes Congress as the legislative branch of federal government, imbued with law-making and funding powers. The Office of the President is the executive branch, empowered to approve or overturn Congressional law and to serve as chief executive of many national activities. The Supreme Court is ultimate authority of the judicial branch, charged with settling any disputes in the interpretation of law.

These constitutional delineations served a younger, smaller, and less complex nation. But as the United States expanded its territories, it became more complicated to govern, and the Congress and president found it expedient to delegate certain powers to regulatory agencies within Cabinet departments in the executive branch.

Regulatory Agencies as the "Fourth Branch" of Government

The Interstate Commerce Act of 1887, created by the 49th Congress and signed by President Grover Cleveland, established the first regulatory agency: the *Interstate Commerce Commission (ICC)*. During this time in American history, railroads were privately owned monopolies expanding westward, and their expansion often conflicted with the interests of farmers and landowners, leading to unrest and agitation. Laws are often created in direct response to public tension, clarifying rules for new forms of social interaction in public life. The development and expansion of railroads called for a governmental response to regulate how this new technological and economic force would impact the American public.

As the first regulatory agency, the ICC would be the model for future agencies to come. These regulatory agencies were often created by Congress or the president to confront and resolve complicated matters arising from the growing pains of an expanding nation. Regulatory agencies are created by acts of Congress, laws which describe their mission and scope, and staffed by technical and industry experts from the sectors of American society over which these agencies hold sway.

Many regulatory agencies were created during the mid-twentieth century to mitigate the devastation and social upheaval created by the Great Depression and World War II. With no presidential term limits, Franklin D. Roosevelt served four terms in office from the end of the Great Depression (1933) to the end of World War II (1945), remaining the longest serving president in American history. The rapid expansion in government spending to counteract the economic collapse during this period led to rapid growth in regulatory agencies during the Roosevelt administration.[4]

To jolt a depressed economy back into life, Roosevelt proposed expansive public work projects, collectively called the New Deal. These projects were managed and overseen by a range of new regulatory agencies, requiring their own set of rules to guide their work. These rules were all codified in a new law passed in 1946, the *Administrative Procedures Act (APA)*.

The APA facilitated the function and growth of government agencies founded during Roosevelt's New Deal, and it guides the work of regulatory agencies to this day. At the heart of the APA is the *"notice and comment" process* allowing direct public participation in how regulations are created. The "notice" is an announcement made about a proposed regulation being considered published in the Federal Register. The "comment" is a 90-day period during which written reactions from the public about the proposed regulation are made. The entire notice and comment process is called *rulemaking* and creates detailed regulations describing how enacted Congressional law is to be applied to the activity about which the law was created.

Regulatory agencies enabled Congress and the president to delegate power to accomplish their legislative goals. Spurred on by the APA and the continuing complexity of American life, regulatory agencies became highly successful in carrying out their missions and are now so firmly embedded within federal infrastructure that they are regarded as a *"fourth branch" of government*.

Two Key Regulatory Agencies: The CDC and FDA

The CDC and FDA are two regulatory agencies within the *Department of Health and Human Services (DHHS)*. DHHS is one of 15 Cabinet departments within the executive branch, one of three "branches of government" established by the United States Constitution. The president leads the executive branch as head of state and appoints all 15 Cabinet department secretaries and the heads of regulatory agencies within these departments, including the CDC Director and FDA Commissioner.

The activities of regulatory agencies like the CDC and FDA are prescribed by specific laws passed by Congress and endorsed by the president. The funding of these activities is made possible by financial appropriations granted by Congress. In fact, every government activity is based on specific laws describing that activity, because the government of the United States is "a government of laws and not of men"[5] and its citizens agree to be governed by "the rule of law."

As the White House COVID-19 Response Team's 2021 booster announcement demonstrates, although the CDC and FDA fall within the large executive Cabinet office of the DHHS, they are not necessarily beholden to executive branch control, having their own influence on American public life. This influence is exercised through pronouncements, industry guidance, and authorizations and licenses granted to vaccine, drug, and device manufacturers.

Legal Framework

The legal framework for how the CDC and FDA operate is based on two laws: the *Federal Food, Drug and Cosmetic Act of 1938 (FFDCA)* and the *Public Health Service Act of 1944 (PHSA)*. These two laws and their amendments guided the work of CDC and FDA throughout the last decades of the twentieth century.

As the United States entered the twenty-first century, existential threats (e.g., the September 11, 2001 terrorist attacks and anthrax bioterrorism events), natural disasters (e.g., Hurricane Katrina in 2005), and infectious disease outbreaks (e.g., SARS in 2002, H1N1 influenza in 2009, and Zika in 2016) ushered in a number of new laws guiding the work of CDC, FDA, and the newly created DHHS Office of the Assistant Secretary for Preparedness and Response (ASPR).

The CDC, FDA, and ASPR play essential roles In responding to chemical, biological, radiological, and nuclear (CBRN) threats made against the United States. Their authorities and functions are defined in five laws passed in the twenty-first century:

- The Public Health Security and Bioterrorism Preparedness and Response Act of 2002
- The Project BioShield Act of 2004
- The Public Readiness and Emergency Preparedness Act of 2006
- The Pandemic and All-Hazards Preparations Act of 2006
- The Pandemic and All-Hazards Preparedness Reauthorization Act of 2013

These laws were enacted in response to disastrous and unforeseen events. A guiding theme throughout these laws is the need for improved cooperation and communication among federal, state, and municipal government entities to real or perceived threats. Specifically, government agency response to CBRN threats must be a well-choreographed set of activities rather than haphazard reactions, all triggered by a declaration of public health emergency.

One expression of this coordination is the 2016 "Playbook for Early Response to High-Consequence Emerging Infectious Disease Threats and Biological Incidents" issued by the Obama administration in response to the Zika outbreak. This playbook specifically describes regulatory agency responsibilities during a CBRN threat, and, though written years before the COVID-19 pandemic, its very existence was a source of controversy during the Trump administration.

Centers for Disease Control and Prevention (CDC)

The CDC[6] is the cornerstone of the United States public health system. Founded in 1946 by epidemiologists working to control malaria, the CDC has continued to practice "shoe leather epidemiology,"[7] solving infectious disease dilemmas in the decades since its inception, including:

- In the 1950s, determining that tainted Salk vaccine was contaminated by a manufacturing process
- In the 1960s, developing new smallpox vaccine administration methods
- In the 1970s, finding the cause of Legionnaire's disease
- In the 1980s, publishing the first case of acquired immunodeficiency syndrome (AIDS) in the June 5, 1981 issue of Morbidity and Mortality Weekly (MMWR)

CDC ORGANIZATION

The CDC is charged with protecting Americans from threats to health, safety, and security. It is organized into five components, each of which focuses on a specific part of the CDC's mission:

- National Institute for Occupational Safety and Health (NIOSH)
- Public Health Service and Implementation Science
- Public Health Science and Surveillance
- Non-infectious Diseases
- Infectious Diseases

During the early months of the COVID-19 pandemic, the CDC advanced understanding of the biology, transmission dynamics, and prevention strategies for the never-before-seen coronavirus SARS-CoV-2, publishing findings on its website and in MMWR, "the voice of the CDC."

FIGURE 6.1
The Roybal campus of CDC in Atlanta, Georgia. (Source: James Gathany/CDC/Public Domain.)

After vaccines were developed and new Delta and Omicron variants emerged, the CDC's COVID Data Tracker, "Interim Clinical Considerations for the Use of COVID-19 Vaccines," Variants & Genomic Surveillance, wastewater studies, and countless guidance materials were indispensable resources for the public's use (see Figure 6.1).

Food and Drug Administration (FDA)

Americans take for granted that food is safe to consume and that medications deemed "safe and effective" by the FDA—whether prescribed by physicians or purchased over-the-counter—will alleviate suffering and not maim or kill them. This was not always the case. The consumer protections afforded by FDA regulatory oversight were actually borne out of three human tragedies.

In 1906, Upton Sinclair published the novel "The Jungle," exposing the underbelly of the meat processing industry. It vividly described the brutal treatment of immigrant laborers working under "wage slavery" and the horrific and unsanitary conditions in which animals were slaughtered in

Chicago's meatpacking plants. It was not only a bestseller, it also caused so much public outrage about meat processing that Congress was compelled to pass the Food and Drugs Act of 1906.

In 1937, sulfanilamide, the first agent shown to be effective against bacteria, was formulated into a pleasant tasting oral solution called "Elixir Sulfanilamide" to treat streptococcal infections. The "elixir" contained a poisonous solvent, diethylene glycol, which is used today in antifreeze. After ingesting "elixir sulfanilamide" 107 people died. Congress was again compelled to act, passing the Federal Food, Drug and Cosmetic Act of 1938 (FFDCA) to promote product safety by prohibiting interstate commerce of any "adulterated or misbranded" food, drug, device, tobacco product, or cosmetic.

By 1962, thalidomide was commonly used to control nausea in pregnant women, mostly in Australia, Canada, Japan, and Europe, where the drug was marketed. After widespread use, some 10,000 children were born with shortened limbs, congenital heart disease, and ear malformations—all birth defects attributed to thalidomide use during pregnancy. These observations led to the Kefauver-Harris Drug Amendments of 1962 to the original 1938 FFDCA.

Today, the FDA oversees close to $3 trillion of food, medicines, and tobacco products consumed by Americans, accounting for about 20 cents of every dollar spent on these products.[8] Its regulatory oversight is performed by nearly 18,000 staff with a $6 billion annual budget. But bringing a new drug to market is time-consuming and expensive. Pharmaceutical manufacturers may require up to 12 years,[9] spending $1.3 billion on average,[10] before receiving FDA approval to bring a new drug to market.

FDA ORGANIZATION

The FDA's mission is to protect public health in the United States, which includes ensuring the safety of food, drugs, and consumer products, as well as regulating tobacco products. It is also charged with providing public education and responding to public health threats. To implement such a broad mandate, the FDA is organized into 14 centers and branches:

- Center for Biologics Evaluation and Research
- Center for Devices and Radiological Health

- Center for Drug Evaluation and Research
- Center for Food Safety and Applied Nutrition
- Center for Tobacco Products
- Center for Veterinary Medicine
- Oncology Center of Excellence
- Office of Regulatory Affairs
- Office of Clinical Policy and Programs
- Office of External Affairs
- Office of Food Policy and Response
- Office of Minority Health and Health Equity
- Office of Operations
- Office of Policy, Legislation, International Affairs
- Office of the Chief Scientist
- Office of Women's Health

FDA Approval Process

Protecting the American public from the possible harms of medications, vaccines, and medical devices is a top priority for the FDA. To ensure the safety and effectiveness of these agents, the FDA has established formal procedures for interacting with pharmaceutical and medical device manufacturers. Before any of these companies can distribute a specific product in the United States, they must obtain an FDA license to market and sell that product.

Often the goals of the FDA and manufacturers are opposed. The FDA is concerned with safety and effectiveness, while manufacturers are eager to get their products to market as soon as possible. Tensions may be heightened for vaccines given to healthy people, where safety and effectiveness are especially important, just as manufacturers are concerned with high costs of bringing these vaccines to market.

To ensure the highest likelihood of safety and effectiveness, the FDA usually requires manufacturers seeking licensure for drugs, vaccines, or medical devices to do the following:

- Submit an Investigational New Drug (IND) application as soon as pre-clinical studies indicate suitability of an agent before clinical trials are conducted on humans

- Obtain written approval from an Institutional Review Board (IRB) about protecting human subjects participating in clinical trials
- Execute Phase 1 through Phase 3 clinical trials to demonstrate safety and effectiveness of the product being studied
- Submit a new drug application (NDA) or Biologic License Application (BLA) to one of six FDA Centers: the Center for Drug Evaluation and Research (drugs), the Center for Biologics Evaluation and Research (vaccines), or the Center for Devices & Radiological Health (medical devices)

Emergency Use Authorization

Among its many activities during the COVID-19 pandemic, the most important and controversial FDA action was relaxing the time-consuming NDA process by allowing *emergency use authorization (EUA)* for several *medical countermeasures (MCM)*, including diagnostic tests, personal protective equipment, therapeutic agents, medical devices, and vaccines.

Several years and millions of dollars are routinely spent by manufacturers to bring new products to market. During a public health emergency requiring the use of medical countermeasures of drugs, vaccines, or medical equipment which may be new, experimental, or not previously licensed, these routine FDA procedures are cumbersome, preventing Americans from accessing life-saving modalities with the haste necessary to counteract rapidly spreading infection.

This became apparent after 9/11 and the anthrax bioterrorism attacks which followed. The demonstrated need to quickly get new drugs or other MCMs to the public led to passage of the Project BioShield Act of 2004, which states as one of its major purposes "streamlining the Food and Drug Administration approval process of countermeasures."[11]

This law made provisions for emergency use of products not yet approved or licensed by the FDA, allowing the DHHS Secretary to temporarily allow these products to be used once a public health emergency has been declared. Temporary use under an EUA is only allowed while a declared public health emergency remains in effect.

The Project BioShield Act of 2004 stipulates that after consulting with the Director of NIH and CDC where the Secretary "concludes that an agent specified in a declaration can cause a serious or life-threatening condition," EUA can be employed if the FDA reasonably believes that:

- The product may be effective in diagnosing, treating, or preventing the serious or life-threatening disease or condition
- The known and potential benefits of the product outweigh the known and potential risks
- There is no adequate, approved, and available alternative to the product[12]

After its 2004 enactment, Anthrax Vaccine Adsorbed was granted an EUA January 14, 2005, but this has since been revoked given termination of the anthrax public health emergency during which the EUA was issued. Similarly, during H1N1, MERS, Ebola, and Zika epidemics, EUAs were issued and have since been revoked.

The full measure of EUA was realized during the COVID-19 pandemic. By the end of 2021, FDA had authorized "14 therapeutics for emergency use including antiviral drugs and monoclonal antibody treatments" and had "authorized, approved, granted, or cleared over 2,000 additional COVID-19 medical products including molecular diagnostic, antigen and serology tests, sample collection devices, personal protective equipment and ventilators."[13] As of March 2022, the FDA had also authorized three SARS-CoV-2 vaccines under EUA.

The Pandemic as Civics Lesson

Regulatory agencies are fundamental players in the development and implementation of medical advances. During the pandemic, important developments in medical science arose: safe and effective vaccines, mRNA as a completely new vaccine platform technology, new oral antiviral medications, a range of diagnostic testing approaches, and accessible user-friendly epidemiologic data analytics. Not one of these breakthroughs or developments would have been possible without federal government involvement, especially the technical expertise and oversight of regulatory agencies like the CDC and FDA.

COVID-19, the worst public health disaster to befall the United States since the 1918 influenza pandemic, demonstrated how modern American government works—and the crucial role of regulatory agencies in the government's response to emerging, unforeseen threats to the nation.

Looking Back, Looking Ahead

Regulatory agencies emerged at the end of the nineteenth century in response to new challenges brought about by a growing nation. During the twentieth century, technological and economic shifts brought rapid changes, and more agencies were established to regulate their effect on the fabric of American life. The United States developed a robust preparedness matrix consisting of laws, regulations, and federal agencies to counteract twenty-first century threats.

This ecosystem of agencies and regulations set the stage for rapid response to COVID-19 in the United States, enabling life-saving testing, therapeutics, and vaccines to reach the American public. Yet the process was not without difficulties. By superseding the established process for FDA and CDC review, the White House COVID-19 Response Team ignited confusion with its premature booster plan announcement—and subsequent disagreement between the FDA and CDC only exacerbated public concerns.

Regulatory agencies will play a crucial role in responding to emerging infectious diseases (EIDs) in the future. But competing interests—political tensions within the executive and legislative branches, economic forces affecting producers of medical manufacturers, intra- and inter-agency discord—can complicate the process and instill public confusion. Coordination among agencies and the branches of government is required for successful efforts as public health crises arise in the future.

Of course, regulatory agencies in the United States are only one component of responding to a pandemic-level EID that spans the globe. The interactions among other countries' regulatory bodies and international organizations have profound effects on the course of a pandemic, as we will explore further in Chapter 9: Global Concerns.

FUTURE OUTINGS

To understand how the lessons of COVID-19 can be applied to future EIDs, ongoing excursions into the field must explore:

- The oversight structures that empower or impede regulatory action
- Structures for coordination among regulatory agencies and branches of government

- The implications of EUAs for public trust in government policies
- The factors that drove discord among the FDA, CDC, and executive branch during the United States' response to COVID-19
- The interaction among the FDA, CDC, and regulatory bodies outside the United States

Student Research Questions

1. What factors may have driven the White House COVID-19 Response Team's decision to announce a booster plan before following established procedures for FDA and CDC review? What was the effect of this decision on public perception of the government's response to the pandemic?
2. Investigate the history of regulatory agencies in the United States. Select one and examine the factors that led to its creation, the problems it was intended to solve, and the effect it has had.
3. How does the FDA's approval process for new drugs and medical devices affect public health? What is the impact on public safety? How does it influence the activities of drug manufacturers?
4. What are the pros and cons of EUA in response to public health crises? In the COVID-19 pandemic, how were concerns about safety balanced against the need for rapid action? What has been the effect on scientific development, government policies, and public sentiment?

NOTES

1. August 18, 2021 transcript of the Press Briefing by White House COVID-19 Response Team and Public Health Officials.
2. Lev Facher, September 1, 2021, STAT News, "Biden pledged to 'follow the science.' But experts say he's sometimes fallen short," https://www.statnews.com/2021/09/01/biden-pledged-follow-the-science-but-has-fallen-short/?utm_campaign=rss.
3. CDC Statement on ACIP Booster Recommendations press release, September 24, 2021, https://www.cdc.gov/media/releases/2021/p0924-booster-recommendations-.html.
4. Roosevelt died in office on April 12, 1945 after serving 12 years as President. One political result of his unprecedented four-term presidency was passage of the 22nd Amendment to the Constitution, limiting a president to two terms and not more than ten years in office.

5. This is the concluding sentence of Part the First, Section XXX, Massachusetts Constitution of 1780 attributed to John Adams, second President of the United States. The Massachusetts Constitution is the model for the United States Constitution ratified in 1788.

6. First called the "Communicable Disease Center" at its founding in 1946, its name was changed to "Center for Disease Control" in 1970, and after a major reorganization in 1981 became the "Centers for Disease Control" to which "and Prevention" was added in 1992, with Congress retaining the "CDC" acronym throughout.

7. "Historical Perspectives History of CDC," *Morbidity and Mortality Weekly*, June 28, 1996, volume 45, number 25, pages 526–530, https://www.cdc.gov/mmwr/preview/mmwrhtml/00042732.htm, accessed 9/21/2021.

8. "FDA at a glance," US Food & Drug Administration, Office of the Commissioner, November 2020.

9. Gail Van Norman, "Update to Drugs, Devices, and the FDA: how recent legislative changes have impacted approval of new therapies," *JACC: Basic to Translational Science*, August 2020, volume 5, number 8, pages 831–839, https://doi.org/10.1016/j.jacbts.2020.06.010.

10. Wouters, OJ, et al, "Estimated research and development investment needed to bring a new medicine to market, 2009–2018," *Journal of the American Medical Association*, March 3, 2020, volume 323, number 9, pages 844–853.

11. US Public Law 108-276, Project BioShield Act of 2004, July 21, 2004.

12. Project BioShield Act of 2004.

13. "FDA 2021 Year in Review," https://www.fda.gov/media/155422/download, accessed April 20, 2022.

7

Vaccines

The Field: The moment SARS-CoV-2 was identified, scientists leaped into action—seeking to create a vaccine that could prevent the spread of infection. Within 11 months, the first vaccine was approved for emergency use: by far the most rapid development of a vaccine in history. At the same time, many factors have complicated the development and public acceptance of SARS-CoV-2 vaccines. Fundamental to preventing the spread of emerging infectious diseases (EIDs) is an understanding of how vaccines came to be, how they interact with the immune system to prevent infection, and how a landscape of confusion and resistance affected SARS-CoV-2 vaccines.

Field Sightings: variolation, antigens, antibodies, innate immune system, adaptive immune system, antigen-presenting cells (APCs), T-cells, B-cells, "artificially-induced infections", pathogen-associated molecular patterns (PAMPs), major histocompatibility complexes (MHCs), dendritic cells, cytotoxic T-cells (CD8+ or Tc cells), helper T-cells (CD4+ or Th cells), plasma cells, memory cells, antigenic epitope, adjuvants, whole-pathogen vaccine, subunit vaccine, nucleic acid vaccine, clinical trials, preclinical phases, phase 1 clinical trial, phase 2 clinical trial, phase 3 clinical trial, spike glycoprotein, messenger RNA (mRNA) nanoparticle vaccines, recombinant viral vector vaccines, inactivated SARS-CoV-2 vaccines, Operation Warp Speed, "anti-vax" movement, "vaccine mandates", *Jacobson v Massachusetts*.

FIELD EXPEDITION: JERYL LYNN'S MUMPS

Five-year old Jeryl Lynn Hilleman woke up at 1:00 a.m. on the morning of March 21, 1963, with a bad sore throat. Her father, Dr. Maurice

DOI: 10.1201/9781003310525-8

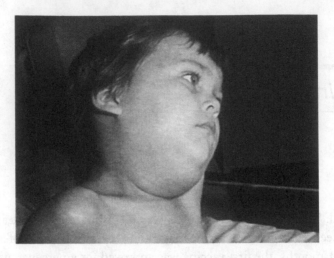

FIGURE 7.1
Parotid gland swelling from mumps virus. (Source: CDC/NIP/Public Domain.)

Hilleman, felt her neck and detected swollen salivary glands, which were characteristic of mumps infection (see Figure 7.1).

During the 1960s, mumps was a common childhood infection, known for causing salivary gland swelling and serious complications of deafness, meningitis, and encephalitis.

Dr. Hilleman was not only a worried father awakened by an acutely ill daughter late at night but also a research scientist at the pharmaceutical company Merck Sharp and Dohme. He decided to swab his daughter's throat, then drove the sample to his lab in the hopes it would help find ways to eradicate the disease in the future. Indeed, he later used this sample to produce the "Jeryl Lynn Strain" of live attenuated mumps virus, and within four years he created the first effective mumps vaccine[1] (see Figure 7.2).

Prior to Dr. Hilleman's development of the mumps vaccine at Merck in 1967, no vaccine had been developed in less than ten years. Creating the mumps vaccine in just four years was unprecedented. However, ten years remained the standard vaccine development timeline for another 53 years. The next major advance in rapid vaccine development occurred on December 2, 2020, when the Medicines and Healthcare Products Regulatory Agency of the United Kingdom authorized the Pfizer-BioNTech messenger RNA-based SARS-CoV-2 vaccine, which was developed in just 11 months using cutting-edge new vaccine platform technology.

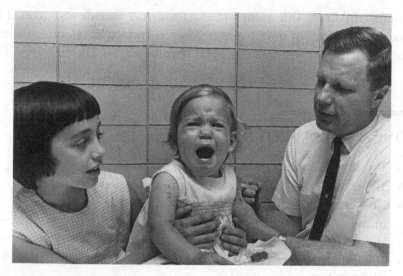

FIGURE 7.2
Kirsten Jeanne Hilleman being vaccinated by Dr. Robert E. Weibel using the investigational Jeryl Lynn strain mumps vaccine, named for her sister, who comforts her. (Source: Gift of the Hilleman Family/National Museum of American History.)

In January 2021, Lorraine Hilleman, Dr. Hilleman's wife, received her first dose of an mRNA SARS-CoV-2 vaccine, observing that its development was "accomplished in an unprecedented short period of time."[2]

LESSON LEARNED: JERYL LYNN'S MUMPS

The substantially shortened timeline for successfully developing a safe and effective SARS-CoV-2 vaccine, from the usual ten years to as little as eleven months, was a remarkable scientific achievement. The messenger RNA technology used to develop the Pfizer-BioNTech vaccine was the first time this new platform for vaccine development was ever used, and it would prove to be a powerful tool in vaccine development as the COVID-19 pandemic continued.

But rather than celebrate the achievement of having a safe and effective vaccine against SARS-CoV-2 finally available—as millions of parents did in the 1960s when the first mumps, measles, rubella, and polio vaccines were finally made available to prevent illness, disability, and death from these common childhood infections—the SARS-CoV-2 vaccine was greeted by some with skepticism or outright resistance. In fact, development speed may have actually contributed to public cynicism.

CONCEPTS COVERED IN THIS CHAPTER

- The advent of variolation and vaccines
- The eradication of smallpox through vaccination programs
- How vaccines work
- Immune system function and response to infection
- How vaccines are formulated
- The import and process of clinical trials
- SARS-CoV-2 vaccine development
- Confusion surrounding SARS-CoV-2 vaccines
- Vaccine hesitancy and resistance

Dr. Jenner's Variolation Experiments

Vaccines deliver non-infecting portions of a pathogen to produce an immune response. Though considered modern achievements, the basic principles of how vaccines work were first demonstrated over 200 years ago by the smallpox variolation experiments of Dr. Edward Jenner.[3]

Smallpox is now an unfamiliar disease, but for many centuries it plagued human populations, causing a severely disfiguring skin condition, blindness, and death (see Figure 7.3).

In the late 1700s, Dr. Jenner, a British physician, knew that cowpox was a skin abnormality common among milkmaids—an annoying nuisance on the hands, but nothing more. Cowpox lesions, called variola, were similar in appearance to smallpox lesions, but milkmaids having cowpox variola did not seem to develop smallpox disease.

This observation became the basis for Jenner's *variolation* experiment. He scraped cowpox pus from a milkmaid, applied it to the wound of a boy, and observed that the variolated boy did not develop smallpox when exposed to smallpox pus.

In conducting his variolation experiment, Jenner did not know that cowpox pus in milkmaid variola contained a cowpox virus that was similar

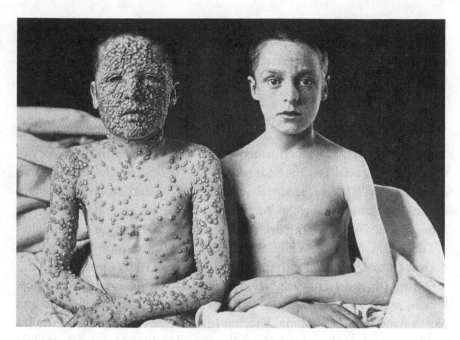

FIGURE 7.3
Disfiguring smallpox in two boys. (Source: Allan Warner/Public Domain.)

enough to the smallpox virus to engender a protective immune reaction in a variolated boy but different enough to not cause actual smallpox disease. By using cowpox pus in this way, Jenner created and successfully used the first early form of a vaccine.

A Vaccination Triumph: The Eradication of Smallpox

There is no single infectious illness that has played a more important role in the history of vaccination than smallpox. While its origin is unknown, it is believed to have been present in humans since the fourth century, spreading around the globe as civilizations expanded and trading routes brought more populations into contact with each other. By the eighteenth century, it was present on every continent on earth.

Smallpox was not only highly infections, it was incredibly deadly. Between 30% and 35% of those infected usually died, with death rates soaring as high as 98% during particularly lethal regional outbreaks, and no effective treatments existed. Those who survived were often badly scarred or blinded. The only protection against smallpox was to avoid contact with infected persons—often impossible for populations in a region experiencing an outbreak—or to contract it and survive. In fact, people who had recovered from smallpox were often called upon to tend to the ill, as they did not contract the disease a second time.

The Smallpox Vaccine

Dr. Jenner's variolation experiments opened a new chapter in the history of this disease. He published his findings in a 1798 booklet titled *An Inquiry into the Causes and Effects of the Variolae Vaccinae, a disease discovered in some of the western counties of England, particularly Gloucestershire and Known by the Name of Cow Pox*, which began circulating among the medical communities of Europe and America. While it inspired a good deal of controversy, several doctors began to implement the practice and share it with others.

During the nineteenth century, Dr. Jenner's variolation process—now known as vaccination—was in widespread use in much of Europe and North America. The National Vaccine Institute was established in the US in 1813 to prevent the spread of smallpox, and in 1855 Massachusetts was the first state to require children to be vaccinated before attending school. Ongoing medical research and public health initiatives allowed vaccine programs to flourish over the next century, and the disease was eliminated in North America and Europe by 1952 and 1953, respectively.

In other parts of the world, however, smallpox was still widespread. The World Health Organization (WHO) launched a global eradication campaign in 1959, then increased its efforts with the Intensified Eradication Program of 1967. This enabled the production of more vaccine doses, intensive case surveillance, and successful mass vaccination campaigns. The last known case of the disease occurred in 1978, and on May 8, 1980, the WHO declared smallpox to be officially eradicated from human populations around the globe. Almost 200 years after Dr. Jenner conducted his first variolation experiment, humankind had for the first time fully triumphed over a virus—thanks to vaccination.

How Vaccines Work

While Dr. Jenner's experiment didn't reveal how variolation with cowpox was able to prevent smallpox infection, the fact that it worked opened the door to research on the mechanisms behind vaccination. How, then, does cowpox pus act as a vaccine?

Cowpox pus contains cowpox virus, which is detected by the body's immune system as a foreign invader. Viruses and all other foreign invaders—bacteria, fungi, and parasites—are pathogens that can cause disease. These disease-causing pathogens contain molecules called antigens that activate the immune system. Antigens cause the immune system to create **antibodies**; the word "antigen" itself is a contraction of "**anti**body **gen**erator."

In Dr. Jenner's experiment, the antibodies generated by the host cells of the variolated boy were targeted against cowpox virus. But these same antibodies were also able to attack smallpox virus because it is genetically very similar to cowpox virus. Therefore, when the boy was exposed to the smallpox virus, the antibodies recognized and attacked the antigen, preventing smallpox disease from developing.

All vaccines we know today contain antigens, which activate the immune system to create antibodies. These antibodies perform two functions: an immediate "direct attack role," which controls acute infection, and a longer term "memory role" that remembers the antigen, allowing the immune system to fight acute infection should exposure occur again.

Vaccines Usurp the Immune System

People are constantly exposed to antigens as they move through the world, and the immune system responds accordingly. A vaccine is a manmade tool that instigates the immune system response to create antibodies to the antigens formulated into the vaccine, which helps the body avoid infection when exposed to those antigens in the future.

Essentially, vaccines trick the body into thinking an actual infection has occurred, when in fact it has not—the body has simply been exposed to an antigen in a controlled way that allows for appropriate immune system response to develop antibodies. In this way vaccines usurp the immune system for the good of preventing future infections when people encounter the targeted antigens in the world.

The "immune system" is actually two interrelated systems: the innate immune system and the adaptive immune system.

The *innate immune system* is the first line of defense against foreign invaders. It consists of:

- The skin and mucous membranes, which create physical barriers to infection
- Gastric acid in the stomach, which kills ingested pathogens
- The fever response, which raises body temperature to inhospitable levels for microbial organisms
- Specialized immunologic cells (macrophages, neutrophils, and others) and molecules (cytokines and the complement system as examples), which can be recruited to destroy pathogens

The *adaptive immune system* is the second line of defense against foreign invaders, augmenting the work of the innate immune system. The adaptive immune system is distinguished by interactions among three immunologic cell types: *antigen-presenting cells (APCs), T-cells, and B-cells*. These immunologic cell types are important both in the body's response to "real infections" caused by viruses and other pathogens and *"artificially-induced infections"* caused by vaccines.

Immune System Response to Infection

The human immune system is fine-tuned to recognize pathogenic cells and react to their antigens. Bacteria, fungi, parasites, and viruses are very different from human cells in terms of their cellular wall structure and organization of genetic material. These biological differences create distinct *pathogen-associated molecular patterns (PAMPs)*. The body's immunologic cells easily recognize these PAMP non-human cellular characteristics and respond by targeting the invading cells for destruction and removal.

While the PAMPs of viruses make them recognizable to the immune system, they are harder for the body to detect. Because viruses consist of just two elements—genetic material (RNA or DNA) encased within a transport vehicle (lipid envelope)—they can quickly hide within cells after injecting their genetic material directly into cellular cytoplasm. Their presence within infected cells would be masked from detection if not for the actions of *major histocompatibility complexes (MHCs)* within cells.

MHC structures within host cells detect foreign proteins like viruses, package these foreign substances into fragments, and push these packaged fragments to external surfaces of infected host cell membranes, signaling to the immune system that these cells have been taken over by foreign invaders. These MHC-surface signals, the extruding fragments of invading virus, can be detected by circulating immunologic cells.

Interactions among Immunologic Cells of the Adaptive Immune System

The adaptive immune system and its three immunologic cell types come into play as antigens are recognized (through PAMP structures and MHC-surface fragments) and immunologic cells go into action. A choregraphed sequence of actions instigate a sort of relay race within the immune system in response to the presence of antigens in the body.

Antigens are first detected by circulating APCs. **Dendritic cells** are the most numerous type of APCs. Dendritic cells are specialized lymphocytes with undulating, elongated, branch-like shapes having abundant room to incorporate numerous antigens on their cell surfaces. Dendritic cells capture and carry these incorporated antigens throughout the bloodstream and lymph fluid until arriving at lymph nodes.

Within lymph nodes, dendritic cells "present" their transported antigens to undifferentiated T-lymphocytes, or T-cells, causing these cells to morph into two new forms: **cytotoxic T-cells (CD8+ or Tc cells)** and **helper T-cells (CD4+ or Th cells)**. As their names imply, cytotoxic T-cells destroy viral-infected cells and helper T-cells assist antiviral efforts by releasing cytokines (interleukins and interferon), encouraging more T-cells to destroy viral-infected cells.

After dendritic-presented antigens stimulate the creation of new cytotoxic and helper T-cells, the third important immunologic cell type comes into play: B-cells. B-cells are stimulated by helper T-cell actions to differentiate into antibody-producing **plasma cells**, which create antibodies specifically against the dendritic-presented antigen; some converted plasma cells become **memory cells**, also retaining the ability to create specific antibodies in the instance of future invasion by the same antigen.

Vaccine Formulation

Antigens are the main component of vaccines. Well-designed vaccines contain antigens that mimic their corresponding pathogens so closely as to cause the exact immune response of an actual infection without making the vaccinated person sick.

The goal of vaccine development is to find antigens that closely mimic their corresponding pathogens. Historically, this meant creating vaccine antigens by weakening or killing live pathogens through physical, chemical, or radiation means, hopefully retaining the pathogen's immune-activating ability while nullifying the pathogen's disease-causing effects.

However, recent advances in genetic science have allowed vaccine developers to create antigens using the pathogen's own DNA or RNA. This precision hinges on knowing which specific part or parts of a virus activate the immune system to create neutralizing antibodies against the pathogen to which it corresponds; this part of the antigen is called the *antigenic epitope*.

Once identified and isolated, vaccine antigens are formulated into solution (sometimes fortified by *adjuvants*, chemicals which further enhance adaptive immune responses). Once the solution is injected, an artificially-induced infection triggers the same immune responses described above as would real infection by the pathogen.

Types of Vaccines

There are three types of vaccines, and they are categorized by the type of antigen they contain:

- *Whole-pathogen vaccines* contain the infecting pathogen itself, which has been weakened (called "live-attenuated virus vaccine") or killed (called "killed or inactivated vaccine") by physical or chemical means so as to not cause actual infection when injected. The mumps, measles, and rubella vaccine is an example of a live-attenuated virus vaccine.
- *Subunit vaccines* contain parts of the whole pathogen, such as portions of its cell membrane or other protein components. Meningococcal vaccines are examples of this vaccine type.
- *Nucleic acid vaccines* use the pathogen's own DNA or RNA to encode for proteins important for causing infection and instruct human cells to create the antigenic components against which an immune response is generated. SARS-CoV-2 vaccines using messenger RNA technology are examples of nucleic acid vaccines.

Vaccine Clinical Trials

Creating vaccines that mimic an exact immune response without actually causing illness is a complicated and potentially dangerous task.

Vaccines must be safe to give to healthy people, elderly persons, infants and children, pregnant or nursing mothers, and people with chronic medical or immunocompromising conditions. At the same time, vaccines must be effective in creating immune responses that help vaccinated people avoid future infection from the pathogens against which the vaccine is designed.

To reduce the potential risk associated with vaccines and to determine their effectiveness, vaccines are tested within clinical trials.

Clinical Trials and Scientific Truth

Scientific research aims to determine the truth about a matter through objective, verifiable, and reproducible means. Scientific research experiments are one way to determine truth.

One can be reasonably assured that the observed results are true when scientific experiments:

- Are designed to address clear and specific questions
- Do their utmost to avoid prejudice in how experiments are conducted and how results are interpreted
- Can be replicated by other researchers who are able to achieve the same results using the same methods as the original researchers

Clinical trials are scientific experiments using healthy humans as volunteer research subjects and are formally designed to provide objective answers about the safety and effectiveness of vaccines, medications, and medical devices. Findings from clinical trials are often presented by pharmaceutical and medical device manufacturers to government regulators so that their products can be licensed and approved by regulatory agencies for use among the general public.

Before starting a clinical trial, researchers consider the following questions.

- What questions need to be answered about this vaccine, medication, or medical device?
- What endpoints will indicate an answer has been found?

- Which kind of scientific study—case-control study, cohort study, or randomized controlled trial (RTC)—should be used to answer these questions and produce the endpoints of interest?
- How is bias eliminated or minimized as the scientific study is carried out and results are generated?

The answers to these questions determine how a clinical trial is formally designed and executed.

Clinical Trial Phases

Clinical trials are generally conducted in sequential phases.

- *Preclinical phases* occur before any chemical or biological compound is used in human subjects and use either animal or computational models. In these phases, compounds are administered to animals (animal models) or manipulated using computer programs (computational models). Promising agents from preclinical studies are then advanced to human experimentation in three phases.
- *Phase 1 clinical trials* are conducted on a small number of human volunteers to determine the safety profile of the vaccine, medication, or medical device being studied.
- *Phase 2 clinical trials* can be conducted once safety profiles are determined. These trials expand the number of human volunteers, adding hundreds more to determine dosages needed to meet clinical endpoints.
- *Phase 3 clinical trials* expand research further, recruiting thousands of healthy human volunteer research subjects—RTCs are likely to be used at this stage. Should the results of these trials confirm the safety and effectiveness of the vaccine, medication, or medical device being studied, manufacturers may seek review and approval from a regulatory agency to bring a medical product to market.

All clinical trial phases are time-consuming and expensive, and this is especially true for vaccines. Because vaccines can pose serious potential risks to healthy people—and even more risk to children, the elderly, and others with higher risk of negative health outcomes—more care, precaution, and expense are associated with their study. Vaccine manufacturers bear most of these costs and have no guarantee of being rewarded for their investments.

The inherent economic risks in vaccine development—lead times in studying potential vaccine candidates within animal and computational models, designing and enrolling healthy human subjects in all phases of clinical trial experiments, the complex process of regulatory agency application, review, and approval—are significant barriers for pharmaceutical manufacturers.

SARS-CoV-2 Vaccines

After the Pfizer-BioNTech SARS-CoV-2 vaccine was authorized in December 2020, vaccine development continued with full force. By March 2022, the WHO had authorized ten SARS-CoV-2 vaccines for emergency use, as shown in Table 7.1.[4]

TABLE 7.1

SARS-CoV-2 Vaccines Authorized by EMA, FDA, MHRA, WHO by March 16, 2022

Brand name	Manufacturer	Vaccine type	Approval agency
Comirnaty	Pfizer-BioNTech	mRNA nanoparticle	EMA, FDA, MHRA, WHO
Spikevax	Moderna	mRNA nanoparticle	EMA, FDA, MHRA, WHO
Janssen COVID-19	Janssen Biotech	recombinant viral vector	EMA, FDA, MHRA, WHO
Vaxzevria	AstraZeneca	recombinant viral vector	EMA, MHRA, WHO
Nuvaxovid	Novavax	recombinant viral vector	EMA, MHRA, WHO
Covishield	Serum Institute of India	recombinant viral vector	WHO
Covovax	Serum Institute of India	recombinant nanoparticle	WHO
BBIP-CorV	Sinopharm	inactivated SARS-CoV-2	WHO China, Bahrain, UAE
CoronaVac	Sinovac	inactivated SARS-CoV-2	WHO China
Covaxin	Bharat Biotech	inactivated SARS-CoV-2	WHO

(*Sources:* European Medicines Agency; Centers for Disease Control and Prevention; Medicines and Health Products Regulatory Agency; World Health Organization.)

In the United States, the Food and Drug Administration (FDA) had only authorized three of the ten WHO-approved vaccines. The regulatory agencies for the European Union (European Medicines Agency, or EMA) and United Kingdom (Medicines and Health Products Regulatory Agency, or MHRA) had authorized five.

The ten authorized SARS-CoV-2 vaccines listed fall into three categories: messenger RNA (mRNA) nanoparticle vaccines, recombinant viral vector vaccines, and inactivated SARS-CoV-2 vaccines.

The mRNA and recombinant viral vector vaccines both employ genetic engineering. They are both nucleic acid vaccines, which use the RNA genetic material that codes for the *spike glycoprotein* (the antigenic epitope of the SARS-CoV-2 virus) to create antigens.

The *messenger RNA (mRNA) nanoparticle vaccines* developed by Pfizer and Moderna include a snippet cut directly from the SARS-CoV-2 genome coding for spike glycoprotein, enclosing this snippet in a lipid nanoparticle. When this formulation is injected into muscle, mRNA within the lipid nanoparticle is incorporated into human cells, instructing these cells to make copies of spike glycoprotein and nothing more. These spike glycoproteins are detected as foreign invaders by MHC complexes or directly extruded into the bloodstream as free-floating spike glycoprotein. In either case, they will be detected by circulating dendritic cells and other APCs, activating T-cells and B-cells to make neutralizing antibodies and memory cells.

The *recombinant viral vector vaccines* developed by Janssen and AstraZeneca use a similar snippet cut directly from the SARS-CoV-2 genome coding for spike glycoprotein, but instead of enclosing this snippet in a lipid nanoparticle, these vaccines splice the snippet into the genome of chimpanzee adenovirus (AstraZeneca) or inactivated adenovirus (Janssen). The adenovirus becomes the transport vehicle for the spliced SARS-CoV-2 spike glycoprotein gene, and this complex is detected by circulating dendritic cells and other APCs, activating T-cells and B-cells to make neutralizing antibodies and memory cells.

The *inactivated SARS-CoV-2 vaccines* developed by Sinopharm, Sinovac, and Bharat Biotech are whole-pathogen vaccines. They use well-known vaccine platform technology with chemically weakened SARS-CoV-2 virus, containing the antigenic properties necessary to induce an immune response, as the basis for their vaccine formulations.

Four Reasons for Rapid Development of mRNA-Based Vaccines

The unprecedented speed with which SARS-CoV-2 vaccines were developed was spurred by the pressing need to curb the transmission of a virus that was itself moving with tremendous rapidity, having spread wide enough to be declared a global pandemic by the WHO within three months of the first cases. How did the scientific community step up to this challenge so quickly? Four factors played a key part: a body of research of into the new technology of an mRNA vaccine platform, efforts begun during recent epidemics, speedy genomic sequencing, and government initiatives.

Earlier Work with mRNA Technology

Use of mRNA technology for therapeutic purposes dates to the 1990s. The power of mRNA technology as a vaccine platform—speed and ease of vaccine design, flexibility and adaptability to antigenic epitope changes, ability to ramp up manufacturing—supports more rapid vaccine development. While challenges are posed by the instability of mRNA-based vaccines, which require ultra-frozen to -70-degree temperatures, a significant body of work already existed to support the creation of an mRNA vaccine against SARS-CoV-2.

Lessons Learned from SARS and MERS

During the viral epidemics of SARS (November 2002–July 2003) and MERS (2012–2014), both limited coronavirus epidemics, vaccine development efforts produced several important revelations related to developing vaccines for coronaviruses. Research revealed these pathogens' tropism (affinity) for the ACE2 receptor, pathophysiology, and disease manifestations. Findings also suggested that spike glycoprotein could be a promising antigenic epitope target for vaccine development, as well as specific antiviral agents that could be used (protease inhibitors containing ritonavir as a booster; remdesevir). Due to the short duration of the SARS and MERS epidemics, vaccine development was aborted, but lessons learned from SARS and MERS augmented creation of SARs-CoV-2 vaccines.

Release of the SARS-CoV-2 Genomic Sequence

Discovering the SARS-CoV-2 genomic sequence was a crucial piece of the puzzle in vaccine development, and it happened quickly. One month after initial cases arose, China released a draft of the genome sequence for "2019 novel coronavirus" on January 10, 2020. This was followed within a few weeks by the Wuhan Institute of Virology's (WIV) complete sequencing of the SARS-CoV-2 genome, which was published on February 3, 2020. With the genomic sequence in hand, Moderna and BioNTech were able to immediately develop mRNA vaccine prototypes, with Moderna beginning human trials on March 16, 2020.

Operation Warp Speed

On May 15, 2020, the Department of Health and Human Services and Department of Defense launched *Operation Warp Speed* to invest billions of dollars to ensure 300 million doses of COVID-19 vaccines would be available by January 2021. This initiative provided significant resources to research and manufacturing, which was aided by the US government pre-purchasing $2 billion of Pfizer vaccine and $483 million of Moderna vaccine, among others. The first successful vaccine candidate was Pfizer's, which was authorized for emergency use by the FDA on December 11, 2020.

Vaccine Confusion: Too Much of a Good Thing

By March 2022, the world was awash in vaccines against the SARS-CoV-2 virus. With ten vaccines in use just two years after COVID-19 was declared a global pandemic, governments and doctors had powerful tools to curb the spread of the virus. In addition, the WHO indicated that 149 vaccines were in clinical development and 195 were in preclinical development.[5]

However, this embarrassment of riches created confusion among scientists, governments, and world populations. Vaccine availability varied widely, as did research and data-sharing, making it difficult to understand the efficacy of different formulations and build on the existing science.

The scientific understanding of how to properly use these vaccines, especially during a time when the SARS-CoV-2 virus was mutating to more transmissible forms like the Omicron variant, continued to change.

The confusion was compounded by continually changing advice about SARS-CoV-2 vaccine use.

- **What does "fully vaccinated" mean?** The Centers for Disease Control (CDC) continued to define a "fully vaccinated" person as someone completing the primary series of a one- or two-dosage vaccine administration, even though numerous studies for the Omicron variant showed waning coverage from the primary series alone, supporting the need for booster doses. Rather than revise its "fully vaccinated" definition, the CDC simply encouraged all Americans to "stay up-to-date with your vaccines," leaving the public to puzzle out what that vague instruction might mean in practice.
- **How much time should elapse between first and second doses?** Young men between 12 and 39 years old were known to develop myocarditis or pericarditis more frequently after receiving first doses of mRNA vaccines, especially the Moderna formulation. After some study of the issue, the CDC recommended extending the time interval between first and second doses of mRNA vaccines from four weeks to eight weeks for this group.
- **Should people be given three doses or four?** A non-peer reviewed study from Israel implied that a fourth dose of the Pfizer vaccine should be given to both high-risk people and 60-year-olds to prevent serious complications or death from SARS-CoV-2 infection.[6]

The result of all these mixed messages was reflected in a January 2022 Pew Research Center survey, in which 60% of respondents were confused about changing public health recommendations for COVID-19, up from 53% in August 2021.[7] Rather than inspire confidence in the efficacy of vaccines, these ongoing changes increased the public's wariness about public health recommendations.

Vaccine Hesitancy and Resistance

In the midst of a global pandemic that had killed six million people around the world as of March 2022, scientists and medical professionals were relieved to have many vaccines at their disposal. Yet, public sentiment did not always mirror this view. From the advent of the first vaccine in December 2020, a vocal anti-vaccine movement expressed resistance to the very mechanism that scientists had so fervently worked to develop.

Existing Background of Distrust

The *"anti-vax" movement* has been particularly visible during the COVID-19 pandemic, but resistance to vaccines isn't new. In fact, in an ironic twist of fate, Dr. Hilleman himself became the target of anti-vaccine rhetoric when a 1998 article in *The Lancet* accused the MMR (measles, mumps, and rubella) vaccine, which he developed based on his 1967 mumps vaccine, of causing autism. The article has since been roundly debunked and retracted by the journal, but the public concern it sparked led to a decrease in MMR vaccination rates and a resurgence in measles outbreaks—which was thought to be eradicated in 2000—that continue to this day.

Vaccination is the most-often performed medical procedure, and the development of vaccines against common lethal pathogens during the twentieth century has saved millions of lives—especially those of children. But many factors contribute to fear, suspicion, and resistance to vaccines, especially in the United States.

- Needle phobia, either as a diagnosed anxiety disorder or a general aversion to needles, dissuades many people of getting a shot at all costs. While reported rates of needle phobia vary widely, it is estimated that one in six adults avoid getting an annual flu shot due to needle phobia.
- Institutionalized racism can drive marginalized populations away from healthcare services, including vaccination. As one example, the history of medical violence perpetrated against African Americans—from J. Marion Sims' experiments on enslaved women in the 1800s to the 1932 Tuskegee Syphilis Study, in which researchers falsely claimed to be treating African American men for syphilis while allowing them to suffer and die from the disease, to the 1951 nonconsensual collection of tissue samples from Henrietta Lacks, from which researchers have profited for decades—has engendered mistrust in the American healthcare system. Inherent racial bias built into the system of care, in general, can dissuade African Americans and other people of color from trusting medical providers; other forms

of bias create similar barriers for transgender people, people of size, neurodivergent people, and other marginalized populations.

- Recency bias is a type of logical fallacy that assigns greater importance to the most recent experiences over historical information, and the public perception of vaccines has shifted in the twenty-first century due to this fallacy. The advent of vaccines for common childhood illnesses such as measles, mumps, and rubella was a relief for parents who had grown up in a world rife with fear about such diseases. Vaccine programs in the twentieth century were incredibly successful—in many cases, nearly wiping out the incidence of these diseases in developed nations—yet this very success set the stage for suspicion about vaccines. Without recent memory of the very real threat of many diseases, public sentiment was often easily swayed by suggestions that vaccines could be the cause of mysterious conditions such as autism, even when those suggestions had no valid evidence.

- Partisan political rhetoric has increasingly targeted vaccines with suspicion. The growth of the anti-vaccination movement during the twenty-first century—accelerated by celebrity activists and the spread of disinformation online—has turned vaccination refusal into a badge of identity for those affiliated with certain political ideologies or conspiracy theories. Some politicians and idealogues have fanned the flames of this movement for their own gain.

Fear, skepticism, distrust, and suspicion regarding vaccines were all in place for many reasons before COVID-19 arose. Coupled with the general fear and disruption to a sense of normalcy that the pandemic created, hesitancy about SARs-CoV-2 vaccines found fertile ground.

State Response to Vaccine Hesitancy

The government has considerable powers to either coerce or persuade citizens to be vaccinated, and *"vaccine mandates"* are not a new phenomenon seen just during the COVID-19 pandemic.

In 1905 the Supreme Court decided in *Jacobson v Massachusetts* that states could exercise their "police powers" during public health emergencies against individual liberties protected by the Due Process clause of the 14th Amendment. This question arose during a smallpox outbreak in Cambridge, Massachusetts, when Henning Jacobson refused to be vaccinated against smallpox, claiming a bad reaction to previous vaccine and incurring a financial penalty for his decision.

During the COVID-19 pandemic, the federal government tested these powers when key public-safety agencies—the Occupational Safety and Health Administration (OSHA) and the Department of Health and Human Services (DHHS)—instituted vaccination rules. The results were mixed: on January 13, 2022, the Supreme Court stayed OSHA's vaccine-or-test rule for employers with 100 or more workers, but upheld the DHHS's mandatory healthcare worker vaccinations for Medicare and Medicaid funded groups and institutions.

At the state level, most governments employed marketing rather than mandates. Campaigns encouraging vaccination were ubiquitous, covering billboards and buses as well as websites and social media. Many states also employed financial incentives (including lottery tickets, gift cards, Krispy Kreme donuts, and free beer) to entice vaccination compliance, raising ethical questions about these practices. The degree to which state and local governments encouraged and enabled people to get vaccinated also varied depending on the political climate of states, counties, and cities.

Looking Back, Looking Ahead

The ideas that led to modern vaccination as we know it today took shape over 200 years ago. Throughout the twentieth century, advances in vaccine science changed the face of infectious disease—and twenty-first century advancements enabled even more sophisticated development. But with modern advancements come modern problems: a confusing, disjointed landscape of vaccine development around the world and myriad factors driving resistance to vaccination.

As new variants of COVID-19 continue to emerge and other EIDs arise, there is a tremendous body of scientific research and emerging technologies that equip researchers with the tools to make prevention possible—and, as the tremendously rapid development of SARs-CoV-2 vaccines demonstrated, to enable swift intervention in the spread of disease.

The challenge of vaccination, however, are not limited to scientific discovery. Without global coordination among governments and regulatory agencies, the world's best defenses against EIDs can become mired in confusion. Without effective methods to overcome individuals' resistance to vaccines, achieving the level of vaccination necessary to eradicate a disease will remain an unattainable goal. The eradication of smallpox through global vaccination efforts was an unprecedented triumph—but it has not yet been repeated with any other disease for which effective vaccines exist. It is therefore the task of all those fighting EIDs to work on multiple fronts in the service of shepherding vaccines from initial laboratory experiments to safe, effective, and widely accepted administration worldwide.

FUTURE OUTINGS

To understand how the lessons of COVID-19 can be applied to future EIDs, ongoing excursions into the field must explore:

- Ways in which the factors that supported rapid development of SARs-CoV-2 vaccines can be further leveraged
- Additional opportunities that mRNA-based vaccines and genetic engineering present for the future of vaccine development
- How political factors impede transparency and data-sharing among nations concurrently developing vaccines
- The role of public health agencies in shaping public behavior in relation to vaccines
- Historical and modern drivers of vaccine resistance, including factors unique to marginalized populations
- Successful methods by which governments can overcome vaccine resistance

Student Research Questions

1. What was the reaction to variolation by scientists and government officials in Europe and the United States during the 1700s? How did the public perceive variolation, and how did perceptions differ based on class and geographic location?
2. What is the history of smallpox disease prior to the development of a vaccine? When and where did it arise, and how widely and rapidly did it spread? How many deaths did it cause?

3. How was the smallpox vaccine distributed? What was the public perception of the vaccine? What programs and measures allowed for its widespread adoption?

4. Why were the efforts to develop vaccines against SARS and MERS abandoned? How did vaccination development against those diseases differ from twentieth-century vaccine development? How did it differ from the process of developing the SARs-CoV-2 vaccine?

5. What programs did countries outside the United States implement to support the development of SARs-CoV-2 vaccines? What role did regulatory agencies play in these programs? How effective were they?

6. On February 28, 1998, *The Lancet* published "Ileal-lymphoid-nodular hyperplasia, non-specific colitis, and pervasive developmental disorder in children," which alleged that the MMR vaccine was associated with the development of gastrointestinal disease and "developmental regression." During the 12 years before it was officially retracted on February 2, 2010, how did MMR vaccination rates change? How did infection rates of measles, mumps, and rubella change? What were the responses from the scientific community?

7. What are the vaccination rates in the United States for MMR, influenza, and SARs-CoV-2 among different marginalized racial and ethnic groups, including African Americans, Asian Americans, Latinx Americans, Pacific Islander Americans, and Indigenous peoples? How do the vaccination rates compare among groups; how do they compare to rates among Caucasians in the United States? Do rates vary by geography? What factors may underlie any differences in vaccination rates across different populations?

8. What incentive programs did states employ to encourage vaccination? How successful were they, and what factors affected the outcomes of these programs?

NOTES

1. Richard Conniff, "A forgotten pioneer of vaccines," *The New York Times*, May 6, 2013.
2. "Dr. Hilleman's spouse receives historic vaccine," Hilleman—A Perilous Quest to Save the World's Children website, News and Events, https://hillemanfilm.com/news/dr-hillemans-spouse-receives-historic-vaccine, accessed March 16, 2022.

3. Riedel, S. "Edward Jenner and the history of smallpox and vaccination," *Proceedings (Baylor University Medical Center)*, 2005, volume 18, number 1, pages 21–25, https://doi.org.10.1080/08998280.2005.11928028, accessed March 18, 2022.

4. WHO Emergency Use List of COVID-19 vaccines, https://extranet.who.int/pqweb/vaccines/vaccinescovid-19-vaccine-eul-issued, accessed March 16, 2022.

5. WHO COVID-19 Landscape of novel coronavirus candidate vaccine development worldwide, https://www.who.int/publications/m/item/draft-landscape-of-covid-19-candidate-vaccines, accessed March 16, 2022.

6. "Protection by 4th dose of BNT162b2 against Omicron in Israel," Yinon M. Bar-On et.al., pre-printed and not peer reviewed version posted February 1, 2022, doi: https//doi.org/10.1101/2022.02.01.22270232.

7. Pew Research Center, February 2022, "Increasing Public Criticism, Confusion Over COVID-19 Response in U.S.," https://www.pewresearch.org/science/2022/02/09/increasing-public-criticism-confusion-over-covid-19-response-in-u-s/, accessed March 18, 2022.

8

Herd Immunity

The Field: The idea of "herd immunity," when enough members of a population develop immunity to a pathogen to stop further transmission, was re-ignited during the pandemic as a public health approach to controlling the spread of COVID-19. Encouraging the spread of disease to stimulate natural immunity was promoted as a strategy by some before vaccines became available. Due to the high rates of illness and death such an approach created, many argued that conferred immunity through vaccination was the more effective route—and this approach became the primary focus once vaccines had been developed.

Field Sightings: herd immunity (population or community immunity), natural immunity, conferred immunity, herd immunity threshold, basic reproduction number (R0 or R-naught), effective reproduction number (Re or Rt), zugzwang, "The Great Barrington Declaration", "The John Snow Memorandum".

FIELD EXPEDITION: SWEDEN'S RISKY EXPERIMENT

Anders Tegnell is Sweden's leading epidemiologist and has an unusual amount of authority. Most public health experts are merely scientific advisors to their political leaders; they are not imbued with policymaking powers. During something as impactful as the COVID-19 pandemic—with its pernicious ability to kill people of all ages, halt economic activity, and upend life at both an individual and international scale—public health experts are generally not empowered to implement national strategic responses.

DOI: 10.1201/9781003310525-9

But Sweden's constitution[1] allows administrative agencies like Tegnell's Public Health Authority to do precisely that: construct a national strategic public health response to contain the spread of illnesses like SARS-CoV-2. And in the first months of the pandemic, Tegnell decided the best way for Sweden to stop viral infection was achieving herd immunity—a large enough percentage of immunity within the population to stop further disease transmission—through community-wide infection.

"That's the way we work in Sweden," explained Tegnell. "Our whole system for communicable disease control is based on voluntary action. You give them the option to do what is best in their lives. That works very well, according to our experience."[2] Based on these cultural values, Sweden instituted no measures to contain the spread of infection, relying on the idea that infection and subsequent recovery among large numbers of people would lead to herd immunity (see Figure 8.1).

After a few months of this laissez-faire approach, Tegnell announced the May 2020 results of its first COVID-19 seroprevalence study[3] showing 7.3% of Stockholm residents with antibodies to the SARS-CoV-2 virus, a

FIGURE 8.1
Anders Tegnell, state epidemiologist of the Public Health Agency of Sweden. (Source: Frankie Fouganthin/Wikimedia Commons/licensed under CC BY-SA 4.0.)

"little lower" than expected "but not remarkably lower, maybe one or a couple of percent," according to Tegnell.[4]

He had hoped for better news. At a time when its Scandinavian neighbors Denmark and Norway were aggressively limiting social gatherings, closing bars and restaurants, sealing off borders, and shutting ski resorts, Sweden's less stringent strategy for containing the spread of SARS-CoV-2 was thought to be more in keeping with its social norms and traditions. At 7.3% seroprevalence, Sweden not only failed to achieve anything close to what was needed for herd immunity but also experienced remarkable levels of SARS-CoV-2 infections and COVID-19 deaths compared to other Scandinavian countries.

LESSON LEARNED: SWEDEN'S RISKY EXPERIMENT

The results of this population-wide experiment convinced Sweden that unchecked infection was not a workable solution to achieving herd immunity. "This is not a disease that can be stopped or eradicated, at least until a working vaccine is produced," Tegnell later said in April 2020. "We have to find long-term solutions that keep the distribution of infections at a decent level."[5]

With a deadly new pathogen running rampant around the world, the need for a strategy that could stop it was pressing. If only enough people could become immune to SARS-CoV-2, transmission could be curtailed or even halted. Without a vaccine, the only path toward this goal was the hope that people infected would recover from the illness and emerge immune to contracting and spreading the disease in the future. But the consequences of such an approach became clear in Sweden: not only was herd immunity impossible with this strategy, but far more people suffered and died as a result.

CONCEPTS COVERED IN THIS CHAPTER

- Origin and definition of herd immunity
- Herd immunity threshold

- Natural and conferred immunity
- Herd immunity prior to SARS-CoV-2 vaccines
- Herd immunity after development of SARS-CoV-2 vaccines

What Is Herd Immunity?

Herd immunity (also known as population or community immunity) is the indirect protection from infection afforded to susceptible individuals residing in a community where large numbers of others have become immune to that infection through exposure (*natural immunity*) or vaccination (*conferred immunity*).[6] When a population has achieved herd immunity, enough individuals are immune to a particular pathogen that the few who are not immune are highly unlikely to become infected—and, if they do, there is virtually no possibility that contagion will spread to large numbers of other individuals.

The term herd immunity was first used in 1918 by the American veterinarian George Potter to describe eliminating "contagious abortion" (spontaneous miscarriage among cattle and sheep), which was a leading threat to livestock. He noted, "abortion disease may be likened to a fire, which, if new fuel is not constantly added, soon dies down. Herd immunity is developed, therefore, by retaining the immune cows, raising the calves, and avoiding the introduction of foreign cattle."[7]

The idea of herd immunity was later applied to human populations in the 1930s when describing influenza, polio, and typhoid epidemics; it arose again in the 1950s with the advent of widespread vaccination programs (see Figure 8.2).

Herd Immunity Threshold

The **herd immunity threshold** is the point at which additional forward transmission of a pathogen within a community of susceptible individuals stops because a sufficient percentage of people have developed immunity through either natural or conferred means.[8]

This threshold is often quoted as an imprecise range between 70% and 90%. It is stated as an imprecise range because it is based on an imprecise concept: *basic reproduction number (R0 or R-naught)*. The "0" or "naught" refers to the first generation of cases during a developing epidemic.

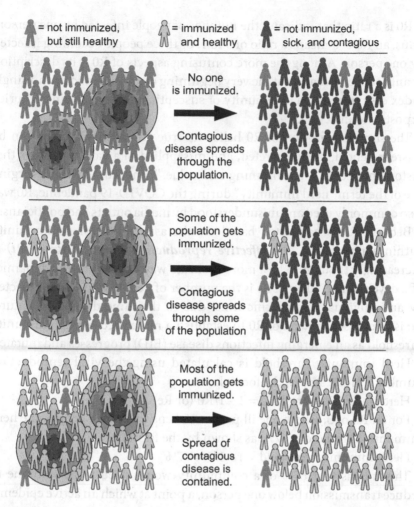

FIGURE 8.2
Spread of a contagious disease is contained as more people are immunized. (Source: Tkarcher/Wikimedia Commons/licensed under CC BY-SA 4.0.)

R0 is the estimated number of healthy people who would be made ill by one infected person. Basic reproduction number was originally used as a demographic term in the 1920s to measure population growth. With the 1957 publication of George Macdonald's seminal work "The Epidemiology and Control of Malaria," basic reproduction number began its epidemiologic use: "daily reproduction rate defined as the expected number of infective mosquito bites that would eventually arise from all the mosquitoes that would bite a single fully infectious person on a single day."[9]

R0 is a ratio that indicates the number of people infected by one person. Thus, an R0 of 3 means a ratio of 3 to 1, or three people becoming infected by one person. Among the more confusing aspects of R0 is its description of an idealized situation at the very beginning of an epidemic, with a single index case infecting a community of susceptible persons having no prior exposure to infection.

The simple definition of R0 belies its true complexity, which "can be misrepresented, misinterpreted, and misapplied in a variety of ways that distort the metric's true meaning and value."[10] Indeed, like the resurging use of the term "herd immunity" during the COVID-19 pandemic, R0 was also commonly used and misunderstood by media outlets and politicians.[11]

Because R0 is calculated based on the assumption of no immunity within a population, the ***effective reproduction number (Re or Rt)*** is increasingly thought to be a more accurate way to describe the dynamics of a developing epidemic. Re is the number of people who can be infected by an individual at a specific time; due to this time-dependent nature, Re is often used instead of R0 to establish a more precise herd immunity threshold as an emerging infections disease (EID) progresses and mutates.

Herd immunity threshold is calculated using the viral reproduction number, as indicated in the formula:

Herd immunity threshold = 1 - 1/R0 (or Re)

For an R0 of 3, where one ill person sickens three well people, the herd immunity threshold is 67%, as shown by the formula:

Herd immunity threshold = 1 - 1/3 = 67%

This means that 67% of a community would need to be immune to reduce transmission below one person, a point at which an active epidemic slows and stops.

Herd immunity threshold for an EID is an ever-shifting target, as ongoing changes to population numbers, infection rates, immunity rates, and pathogen variants alter the landscape.

Herd Immunity during COVID-19

While the idea of herd immunity has been in common use since the middle of the twentieth century, the term took on particular significance—and additional nuance—with the advent of the COVID-19 pandemic. The definition of herd immunity in public discourse shifted during the course of the first two years of the pandemic. Initially, before vaccines became available in December 2020, achieving herd immunity through natural immunity was proposed by some as a public health policy, as illustrated by Sweden's early policy. After vaccines were commonly being used in the United States (from January 2021 onward), the metric for achieving herd immunity through conferred immunity became the explicit target for national vaccination programs.

Natural Immunity

The first use of herd immunity, as a public health policy before vaccines were developed, was similar to what George Potter observed in livestock populations over a century ago. Some members of the scientific community and public health officials proposed that if SARS-CoV-2 was allowed to freely infect human communities "likened to a fire," large numbers of people would develop immunity, and the infection would eventually burn itself out. Pursuing this approach to herd immunity requires accepting the likely outcome of allowing a lethal pathogen to spread unchecked through a susceptible community: many possible deaths and chronic complications from "long COVID" among some who survive infection.

Conferred Immunity

The second use of herd immunity came into use as a public health target after vaccines were developed, based on the metric of herd immunity threshold. In December 2020, as the first FDA-authorized SARS-CoV-2 vaccine from Pfizer-BioNTech was rolled out in vaccination programs across the United States, it was thought that 70% to 90% of a community would need to be fully vaccinated to signal herd immunity had been achieved, which would effectively put a halt to active SARS-CoV-2 transmission as the number of susceptible persons was reduced through vaccination.

Herd Immunity during the Pre-Vaccination Phase

During the pre-vaccination phase, all governments faced the dilemma of responding to a developing crisis thought comparable to the 1918 influenza pandemic, which sickened 500 million (then about one-third of the world's population) and caused 50 million deaths. From the initial identification of the outbreak of 44 cases in January 2020 until September 22, 2020—when 31,374,796 people had been infected and 965,893 were dead worldwide[12]—there were no good options. No vaccines had been developed, and whether developing them was even attainable was uncertain. At the same time, basic facts about the behavior of SARS-CoV-2 were still unknown: how quickly it was transmitted from one person to another, whether infected persons could be asymptomatic carriers of the virus, the elapsed time between viral exposure and active infection. Rapid, impactful decisions had to be made, and doing nothing was itself fraught with deadly consequences.

During this early phase, world leaders were trapped in the chess dilemma known as *zugzwang*. Zugzwang[13] is a German word literally meaning "move" (zug) "obligation" (zwang): the necessity of making a move even if detrimental to the player. As mounting cases of the novel coronavirus quickly spread from China to South Korea and Italy, leaders were obligated to make a move—any move—in this deadly chess game pitting them against SARS-CoV-2. They were all trapped in zugzwang, and something had to be done.[14] It was in this context that the term herd immunity was used to mean population-wide immunity gained through infection and recovery.

Great Barrington Declaration vs. John Snow Memorandum

Even after Sweden's initial experiment produced dismal results, some researchers and policy makers continued to tout natural immunity as a viable path to achieving herd immunity.

One of the last efforts to position herd immunity without vaccines as a viable public health policy occurred on October 4, 2020, just as the pre-vaccination phase was about to come to an end. Three epidemiologists—Dr. Martin Kuldorff of Harvard, Dr. Jay Bhattacharya of Stanford, and Dr. Sunetra Gupta of Oxford—met in Great Barrington, Massachusetts, where they drafted a manifesto entitled *"The Great Barrington Declaration,"* citing "grave concerns about the damaging physical and mental health impacts of the prevailing COVID-19 policies."[15]

This document argued that confinement strategies were the cause of "worsening cardiovascular disease outcomes, fewer cancer screenings, and deteriorating mental health" and that "keeping these measures in place until a vaccine is available will cause irreparable damage." They proposed an ameliorating strategy called "Focused Protection":

> as immunity builds in the population, the risk of infection to all—includ-
> ing the vulnerable—falls. We know that all populations will eventually

reach herd immunity—i.e., the point at which the rate of new infections is stable—and that this can be assisted by (but is not dependent upon) a vaccine. Our goal should therefore be to minimize mortality and social harm until we reach herd immunity. … the most compassionate approach that balances the risks and benefits of reaching herd immunity, is to allow those who are at minimal risk of death to live their lives normally to build up immunity to the virus through natural infection, while better protecting those who are at highest risk. We call this Focused Protection.

Ten days later, on October 14, 2020, another group of scientists—including Dr. Rochelle Walensky, Chief of Infectious Diseases at Massachusetts General Hospital, who would be appointed as Director of the Centers for Disease Control (CDC) in 2021—published an opposing view to the Great Barrington Declaration: *"The John Snow Memorandum"*, honoring the mid-1800s London physician, considered a founding father of modern epidemiology for his research into cholera as a waterborne illness.[16] Their memorandum confronted the harms of lockdown, which was the central tenet of Barrington:

although lockdowns have been disruptive, substantially affecting mental and physical health, and harming the economy, these effects have often been worse in countries that were not able to use the time during and after lockdown to establish effective pandemic control systems.[17]

The John Snow Memorandum also directly addressed herd immunity achieved through natural immunity:

the arrival of a second wave and the realization of the challenges ahead has led to renewed interest in a so-called herd immunity approach, which suggests allowing a large uncontrolled outbreak in the low-risk population while protecting the vulnerable. Proponents suggest this would lead to the development of infection-acquired population immunity in the low-risk population, which will eventually protect the vulnerable. This is a dangerous fallacy unsupported by scientific evidence.

The many scientific signatories to the Great Barrington Declaration and the John Snow Memorandum had no idea that within one month, Pfizer-BioNTech would announce 95% efficacy results from the Phase 3 clinical trial of its messenger RNA vaccine or that, by mid-December 2020, two mRNA vaccines would begin inoculating millions of Americans. Before the advent of these vaccines, there were very few available options other than the "hammer and dance" of lockdowns and resurgence. Zugzwang, the obligation to do something even if it produced harm, meant choosing between two options, each of which produced a different type of harm: (i) employing lockdowns to reduce infections at the expense of economic health and to the detriment of mental health; or (ii) considering herd immunity as a public health policy though unchecked transmission on a national level would lead to higher levels of death and disability from COVID-19 infections.

Herd Immunity during the Post-Vaccination Period

Until the twenty-first century, coronaviruses were thought to be fairly benign, causing nothing more than the common cold. The first Severe Acute Respiratory Syndrome (SARS) epidemic in 2002 and the Middle Eastern Respiratory Syndrome (MERS) in 2012—both caused by coronaviruses—changed how these types of pathogens were regarded. With the appearance of coronaviruses that caused severe illness and death came a new focus on developing vaccines for these types of pathogens.

Vaccine development is an arduous, convoluted process which takes many years, with no guarantee of success. The Ebola vaccine took 43 years to develop from the first Ebola outbreaks reported in 1976 until vaccine approval in 2019. Acquired Immune Deficiency Syndrome (AIDS) caused by human immunodeficiency virus (HIV) was first reported in 1981, and an effective vaccine is still not available. Vaccine development attempts against SARS and MERS were started and abandoned due to the brevity of both epidemics.

Given this track record, skepticism about the prospects of developing a vaccine against SARS-CoV-2 was warranted. At the time COVID-19 was declared a global pandemic by the World Health Organization (WHO) in March, 2020, the most optimistic projected COVID-19 vaccine availability was two years.

Yet despite the odds, safe and effective COVID-19 vaccines were developed and authorized in less than one year. The world entered a post-vaccination phase from December 2020 onward, and the term "herd immunity" took on a new meaning. Conferred immunity through vaccination replaced the idea of natural immunity acquired through infection, and herd immunity threshold became the de facto target for vaccination programs.

Herd Immunity Threshold in the Post-Vaccination Period

Two months prior to the approval of the first SARS-CoV-2 vaccine, the WHO Director-General spoke out against the strategy of unchecked infection as a means of achieving herd immunity in an October 2020 media briefing, stating definitively that "herd immunity is a concept used for vaccination, in which a population can be protected from a certain virus if a threshold of vaccination is reached."

What exactly that threshold should be for COVID-19, however, remained unclear. Thresholds for other diseases had been established—95% for measles and 80% for polio, for example—but even with a vaccine in hand as of December 2020, too little was known about the SARS-CoV-2 virus to establish a threshold for this new disease.

In January 2021, the WHO established the Independent Allocation of Vaccines Group (IAVG) as part of its initiative to promote the fair and equitable allocation of COVID-19 vaccines worldwide. On December 23, 2021, the IAVG issued their "Strategy to Achieve Global COVID-19 Vaccination by mid-2022," setting a goal of vaccinating 70% of the world's population by June of 2022.

The IAVGs 70% vaccination target did not claim to represent a full herd immunity threshold; the stated goal was to achieve "broad vaccination coverage" in order to "minimize deaths, severe disease and overall disease burden; curtail the health system impact; fully resume socio-economic activity; and reduce the risk of new variants." While the target for broad coverage was rooted in ongoing research into COVID-19 epidemiology, it also took into account the feasibility of vaccine production and delivery, along with other global concerns regarding economic and political factors impacting vaccine distribution.

A 70% fully vaccinated goal for the world's population would not be nearly enough to provide the kind of "herd immunity" needed to damp down ongoing transmission of the SARS-CoV-2 virus, especially the Delta variant, which was estimated to have a transmission rate or "reproduction number" ranging from 5 to 9.5.[18] Assuming a reproduction number of 9.5, at least 90% of people would need to be fully vaccinated to stop ongoing SARS-CoV-2 transmission, bringing the pandemic to an end.

The IAVG's vaccination target was believed to provide, at least, a minimum threshold viable enough to mobilize resources needed to close inequities in vaccine manufacturing and distribution, hopefully bringing the world closer to the end of the pandemic. Further research would be needed to establish a definitive herd immunity threshold sufficient to fully stop forward transmission of COVID-19.

Herd Immunity vs. "A Path to Normality"

One of the most highly regarded COVID-19 forecast modelers was Youyang Gu, an MIT graduate with degrees in electrical engineering and computer science, who launched the website Covid-19projections.com in the spring of 2020, which garnered millions of daily views on a webpage entitled "Path to Herd Immunity."

As a data scientist with expertise in finance and sports analysis, Gu had no formal background in epidemiology or public health but became intrigued with COVID-19. He developed a model to forecast infections, deaths, and pandemic recovery timelines. A few months after launching the website, Gu indicated a shift in his thinking through a series of Twitter posts in February 2021, stating that "our goal should not be to reach 'herd immunity,' but to reduce COVID-19 deaths & hospitalizations so that life can return to normal." As a result, he changed the webpage name from "Path to Herd Immunity" to "Path to Normality."

Similar shifts in herd immunity as a vaccination goal were observed throughout the rest of 2021 as vaccination rates fell and the Delta variant waned. Herd immunity began to be viewed less as a hard target and more as a mathematical asymptote, a theoretical endpoint to be approached but not reached.

Looking Back, Looking Ahead

The idea of achieving population-wide immunity to a pathogen arose in the early twentieth century; mid-twentieth century vaccination programs popularized the idea of herd immunity as they sought to eradicate diseases that had long plagued human populations, especially children.

The COVID-19 pandemic re-ignited interest in herd immunity; in the face of a new pathogen against which the world had no defenses, it seemed to many the only hope for an eventual end to a rapidly spreading disease that crossed borders around the world. Prior to the development of SARS-CoV-2 vaccines, the only path to herd immunity hinged on encouraging widespread transmission, which had devastating consequences for public health. But with vaccines in hand, the promise of herd immunity without additional sickness and death provided a possible path toward disease prevention through conferred immunity worldwide.

Yet the threshold for herd immunity remains uncertain. New disease variants and indeterminate vaccine effectiveness complicate the calculation. Even if a threshold is determined, multiple factors, discussed in Chapter 9: Global Concerns, introduce questions about the feasibility of achieving it.

Infectious disease outbreaks have indefinable starting and ending points. Though based on changing dynamics throughout an ongoing outbreak, achieving a quantifiable herd immunity threshold through vaccination programs or infection and recovery could be considered as a definable end point marking the close of a pandemic. Herd immunity, and how it was conceived before and after the development of SARS-CoV-2 vaccines, is the nexus for important pandemic-related questions as the world confronts future EIDs.

Whether a herd immunity threshold for an EID can be conclusively determined may be less important than reaching a reasonable level of population immunity such that disease transmission is low enough to allow for normal societal functioning and a sense of closure for the public. COVID-19 and future pandemic-level EIDs may progress less like smallpox—which was eventually eradicated entirely—and more like the 1918 influenza pandemic: instead of an eradication of all influenza viruses, we live with a disease for which annual vaccination can provide a high degree of protection and the rates of severe illness and fatality are low enough to be deemed acceptable by the public.

FUTURE OUTINGS

To understand how the lessons of COVID-19 can be applied to future EIDs, ongoing excursions into the field must explore:

- How and why did herd immunity become a public policy option in response to COVID-19?
- What were the costs of efforts to achieve herd immunity through the spread of infection? In the absence of vaccines, why was this approach deemed viable by some scientists and officials?

- Are COVID-19 recovered persons sufficiently immune to provide the kind of indirect protection needed by susceptible persons envisioned by herd immunity? How can this be effectively measured in future EIDs?
- What happens to herd immunity threshold calculations as disease transmission changes through viral mutation?

Student Research Questions

1. Explore the history of herd immunity as a public policy in the twentieth and twenty-first centuries. When and how was it employed in relation to large-scale vaccination programs? How was it regarded by the scientific community? How was it received by the public?

2. What herd immunity thresholds have been determined for infectious diseases in the past? How were those thresholds determined, and how accurate have they proved to be?

3. How were the terms "herd immunity" and "reproduction number" used in public discourse during the first two years of the COVID-19 pandemic? How accurate were the definitions provided by public officials? How did the use of these terms change before and after the development of SARS-CoV-2 vaccines?

4. What has been the result of pursuing herd immunity through natural immunity—disease infection and recovery—for various infectious diseases prior to 2020? Where, when, and how was this approach taken in response to the COVID-19 pandemic, and what were the results? How do the outcomes of herd immunity through natural immunity compare to those achieved through conferred immunity via vaccination?

5. When faced with the dilemma of zugzwang prior to the development of SARS-CoV-2 vaccines, what options did world leaders have for slowing or stopping the spread of disease? What approaches did various countries take, and what were the outcomes?

6. How does a policy of pursuing herd immunity affect high-risk populations, including children, the elderly, and those at risk of serious disease? What are the differences in risk and outcome between natural and conferred immunity approaches for these populations?

7. What are the benefits and shortcomings of calculating herd immunity threshold using basic reproduction number? How do they compare to calculations based on effective basic reproduction number?

NOTES

1. Chapter 12, Article 2 "Independence of administration" of the Swedish constitution, the Regeringsform, states: "No public authority, including the Riksdag (Swedish parliament), or decision-making body of any local authority, may determine how an administrative authority shall decide in a particular case relating to the exercise of public authority vis-à-vis an individual or a local authority, or relating to the application of law."
2. "In the coronavirus fight in Scandinavia, Sweden stands apart," C. Anderson, HP Libell, *The New York Times*, March 28, 2020, www.nytimes.com.
3. Seroprevalence studies are population-based surveillance analyses using serologic tests from individual blood samples like antibody titers to detect spreading infection through a community.
4. "Sweden is still nowhere near 'herd immunity'," N. Kennedy, CNN Health, May 21, 2020, www.cnn.com.
5. M. Paterlini, "'Closing borders is ridiculous': the epidemiologist behind Sweden's controversial coronavirus strategy," *Nature*, April 21, 2020, volume 580, page 574, www.nature.com.
6. H.E. Randolph and L.B. Barreiro, "Herd immunity: understanding COVID-19," *Immunity*, volume 52, May 19, 2020, pages 737–741.
7. "Contagious abortion of cattle," Agricultural Experimental Station, Kansas State Agricultural College, Department of Veterinary Medicine, Circular No. 69, August, 1918, page 10. David Jones and Stefan Helmreich, "A history of herd immunity," *The Lancet*, September 19, 2020, pages 810–811.
8. See Note 6.
9. Ross, Macdonald, and a theory for the dynamics and control of mosquito-transmitted pathogens, *PLOS Pathogens*, April 2012, volume 8, issue 4, e1002588.
10. Complexity of the Basic Reproduction Number (R_0), Delamater PL et al., Emerging Infectious Diseases, Centers for Disease Control, volume 25, number 1, January 2019, doi: https://doi.org/10.3201/eid2501.171901.
11. On May 10, 2020, UK Prime Minister Boris Johnson described five "Covid Alert Levels determined primarily by R and the number of coronavirus cases," indicating "everyone will have a role to play in keeping the R down" and that, as of the date of his May 10 statement, "we have the R below one, between 0.5 and 0.9," which seemed to indicate the pandemic's end purely based on social distancing and other public health measures before vaccines were in use.
12. Johns Hopkins University Center for Systems Science and Engineering COVID-19 Dashboard data as of September 22, 2020, 11:23 AM.

13. *Zugzwang* is defined as "a position in which a player must move but cannot do so without disadvantage; the obligation to make a move even when disadvantageous." *Shorter Oxford English Dictionary*, Oxford University Press, sixth edition, 2007, volume 2, page 3710.
14. COVID-19 zugzwang: potential public health moves towards population (herd) immunity, Bhopal RS, Public Health in Practice 1 (July 15,2020), 10031, https://doi.org/10.1016/j.puhip.2020.10031.
15. The Great Barrington Declaration, October 4, 2020, accessed at https://gbdeclaration.org on April 11, 2021.
16. The lesson of John Snow and the Broad Street Pump, Mitali Banerjee Ruths, *Virtual Mentor*, June 2009, volume 11, number 6, pages 470–472, www.virtualmentor.org.
17. Scientific consensus on the COVID-19 pandemic: we need to act now, Alwan NA et al., *The Lancet*, October 31, 2020, volume 396, pages e71–e72, published online October 14, 2020 https://doi.org/10.1016/S0140-6736(20)32153-X.
18. Emily Anthes, "The Delta Variant: What Scientists Know", *The New York Times*, June 22, 2021, https://www.nytimes.com/2021/06/22/health/delta-variant-covid.html.

9

Global Concerns

The Field: Pathogens know no boundaries. In the interconnected global economy of the twenty-first century, effective vaccination rates must be met worldwide to stem the spread of infectious diseases. Resource disparities among high-income and low- and middle-income countries present a unique challenge. Building on twentieth-century vaccine efforts, new public–private partnerships have arisen to tackle modern problems— and may pave the way for future responses to emerging infectious diseases.

Field Sightings: high-income countries (HICs), low- and middle-income countries (LMICs), Expanded Programme on Immunization (EPI), public–private partnerships, Bill and Melinda Gates Foundation, Gavi, Coalition for Epidemic Preparedness Innovation (CEPI), Access to COVID-19 Tools Accelerator (ACT-A), COVID-19 Vaccines Global Access Facility (COVAX).

FIELD EXPEDITION: THE WORLD HEALTH ORGANIZATION STRATEGY

"The pandemic is a long way from over, and it will not be over anywhere until it's over everywhere."

—WHO Director-General Tedros Adhanom Ghebreyesus, opening remarks during his COVID-19 media briefing, May 17, 2021

On October 7, 2021, ten months after global vaccination efforts had begun, the World Health Organization (WHO) released its "Strategy to Achieve

DOI: 10.1201/9781003310525-10

Global Covid-19 Vaccination by mid-2022," calling for all nations to fully vaccinate 40% of their populations by the end of 2021 and 70% by mid-2022.[1]

A strategy to vaccinate 70% of some eight billion people using a two-dose vaccine series meant 11 billion doses would be needed. With 1.5 billion doses produced monthly, the 11 billion dose vaccine production goal could theoretically be met within seven months.

But from the onset of its announcement, the WHO strategy seemed to be in trouble:

- *African nations had low vaccination rates*: most wealthy nations had already well surpassed the WHO strategy's 40% intermediate goal by the end of October 2021, but nearly the entire African continent was well below that mark. Nigeria—the most populous nation in Africa, with 206 million people—had only 2.6% of its citizens partly or fully vaccinated; Ethiopia—the second-most populous, with 115 million people—had only 2.8% partly or fully vaccinated[2] (see Figure 9.1).
- *Problems with pediatric COVID-19 vaccinations*: Children under the age of 14 make up 25% of the world's population.[3] In the US, 14% of the total population is composed of children under 12 years old.[4] The FDA granted Emergency Use Authorization of the Pfizer-BioNTech

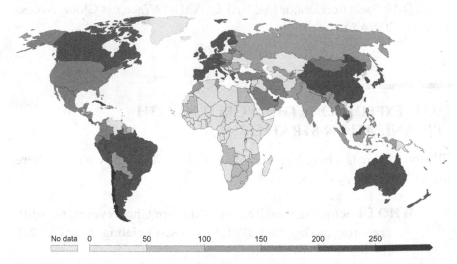

FIGURE 9.1
All African nations have suboptimal vaccination rates compared to rest of world. (Source: Official data collated by Our World in Data—Last updated May 4, 2022.)

vaccine for children five to eleven years of age in November 2021. However, an October 2021 Kaiser Family Foundation survey indicated that 30% of US parents with five- to eleven-year-olds were not planning to vaccinate their children when the vaccine became available,[5] and by February 2022 clinical trials studying COVID-19 vaccines in infants younger than four years old had been halted.

- ***Insufficient goal***: The WHO strategy called for 70% of all nations to be fully vaccinated by mid-2022, but this target would not be nearly enough to provide the kind of population-based immunity needed to damp down ongoing transmission of the SARS-CoV-2 virus, especially the more transmissible Omicron variant that emerged in November 2021.

As the mid-year 2022 deadline approached, what progress was being made to achieve the WHO strategy's 70% fully vaccination goal? By March 12, 2022, 74% of people in ***high-income countries (HICs)*** and 48% of those in ***low- and middle-income countries (LMICs)*** were fully vaccinated.[6] African countries continued to be among the lowest vaccinated nations in the world, with its three most populous nations still having low fully vaccinated rates: Nigeria at 4%, Ethiopia at 16%, and Egypt at 29%.[7] At this rate, meeting the goal of 70% worldwide vaccination would likely be out of reach.

LESSON LEARNED: THE WHO STRATEGY

The WHO strategy brought into stark relief the growing rift between rich countries resourced with vaccine technology and development, manufacturing capacity, and distribution capabilities and poor countries lacking many of these resources. As the gulf widened, SARS-CoV-2 continued infecting human populations, setting the stage for the creation of more transmissible variants like Delta and Omicron.

Wealthy nations assume a degree of security as a benefit of their well-resourced vaccination programs, but this security is threatened when more transmissible variants spawned within unvaccinated poor nations circulate and infect previously vaccinated populations. Continuing infection would make the goal of population-based immunity difficult to achieve, and the WHO Director-General's observation that the pandemic "will not be over anywhere until it's over everywhere" would prove correct.

CONCEPTS COVERED IN THIS CHAPTER

- Global interdependency and the problems of inequity
- The six tasks of global vaccination
- Twenty-first century vaccination challenges
- Twentieth-century vaccination efforts
- Public–private partnerships

What Vaccinating the World Means

There are eight billion people living within 195 nations on planet Earth—countries with different geographies, economies, literacy rates, and political frameworks. In order to gain worldwide population-based immunity, global vaccination efforts must accomplish six tasks:

- **Vaccine availability**: developing and manufacturing a safe and effective vaccine available for widespread use within all age groups, from infants to the elderly, manufactured in sufficient quantities to meet vaccine requirements (single or multiple doses), ready for shipping and distribution
- **Vaccine accessibility**: ability to obtain a sufficient amount of vaccine doses, which is likely to be expensive due to the significant costs associated with vaccine research, development, and manufacturing
- **Vaccine delivery**: getting vaccines to vaccinating sites (clinics, hospitals, physician offices) by modes of transportation (refrigerated trucks, delivery vehicles, and aircraft) equipped to carry vaccines requiring storage at ultra-cold temperatures down to –70 degrees Celsius
- **Vaccine administrators**: trained healthcare workers able to properly mix the vaccine with diluting solution, observe the mixture for consistency, give the reconstituted solution as intramuscular shots directly into the deltoid (shoulder) or gluteal (buttocks) muscle, watch for allergic reactions after giving the shot, document these

tasks, and ensure vaccinated persons return for additional shots if the vaccine is given in more than one dose

- **Vaccine acceptance**: a belief among most of the population that being vaccinated against an illness is better than being unvaccinated and a willingness among individuals to get vaccinated and vaccinate their children

- **Vaccine social equity**: the ability to deliver vaccines to underserved populations, including unhoused persons, those living in rural areas or far from vaccination sites, people who lack the funds or ability to travel to vaccine sites, those without technological means or know-how to access internet-based vaccine scheduling, and people who work jobs that prevent them from visiting vaccine sites during operating hours (especially those at or near the poverty line who work multiple jobs)

Solving these challenges among billions of people who must successfully complete a very specific task—getting one or more intramuscular shots, often in the face of access and availability challenges—is a Herculean undertaking. But the world had experienced success before with vaccination projects aimed at eliminating smallpox and polio. Even some of the logistical challenges of delivering ultra-cold vaccines into politically unstable countries had been overcome during Ebola vaccine campaigns in African nations.

These achievements would need to be remembered and replicated during the global SARS-CoV-2 vaccination campaign.

SARS-CoV-2 Vaccination Begins

Margaret Keenan, a 90-year-old grandmother in the UK, was the first person to be vaccinated against SARS-CoV-2 when she received the Pfizer-BioNTech vaccine on December 8, 2020. The world finally had a powerful technology to stop SARS-CoV-2 infections, and the era of SARS-CoV-2 vaccination came into being.

Having solved the problem of developing a safe and effective SARS-CoV-2 vaccine, global distribution became the next challenge. Two years after the first vaccine administration, HICs had achieved fairly high vaccination rates (74% in March 2022), but LMICs were lagging far behind (48% in March 2022).

The world had confronted vaccine maldistribution challenges before: eradicating smallpox, followed by eliminating the vaccine-preventable childhood illnesses of diphtheria, tetanus, pertussis, polio, and measles in the latter half of the twentieth century.

Valuable lessons were learned over this 50-year period about government commitment to public health activities and the necessity of public–private partnerships to secure funding, transfer intellectual property and technology, and provide technical assistance in healthcare delivery to ensure global vaccination programs succeeded. Would these lessons bear fruit in confronting distribution inequity for the SARS-CoV-2 vaccine?

COVID-19 VACCINATION RATES: MAY 17, 2022

One year after the WHO's vaccination strategy was announced 66% of people in the world had received at least one dose of a COVID-19 vaccine, but vaccination rates varied widely across countries and regions.[8]

- United Arab Emirates: 99%
- Portugal: 96%
- Cuba: 94%
- Chile: 93%
- China: 89%
- Canada: 86%
- Brazil: 86%
- Vietnam: 85%
- Italy: 84%
- Japan: 82%
- France: 81%
- Thailand: 81%
- United Kingdom: 78%
- United States: 78%
- Bangladesh: 78%
- Germany: 77%
- Iran: 76%
- India: 72%
- Indonesia: 72%

- Mexico: 67%
- Pakistan: 60%
- Russia: 56%
- Egypt: 45%
- Ethiopia: 21%
- Nigeria: 13%

Twenty-First Century Problems

Efforts to eliminate smallpox and other childhood infectious diseases provided a template for confronting SARS-CoV-2 vaccine distribution challenges, but the economic landscape of the twenty-first century presented a new set of complexities.

During the first decades of the twenty-first century, the epidemics of SARS (2002), H1N1 influenza (2009), MERS (2012), Ebola (2014,) and Zika (2015) highlighted the financial risk vaccine manufacturers incurred in developing new vaccines to confront emerging infectious diseases (EIDs). Vaccination programs for these EIDs also revealed the lost opportunities resulting from manufacturers abandoning vaccine development efforts once the acute phase of a new epidemic had passed and further underscored the funding and leadership benefits that public–private partnerships afforded for vaccine research, development, and manufacturing.

Twenty-first century outbreaks also made it clear that infectious disease epidemics were not just problems for LMICs. Pathogens have no boundaries, and five SARS-CoV-2 variants of concern began in different locations throughout the world and quickly spread: Alpha (from the UK), Beta (from Brazil), Gamma and Omicron (both from South Africa), and Delta (from India).

A twenty-first century interconnected global economy also means the economies of HICs are severely impacted by economic stagnation within LMIC economies. The International Chamber of Commerce indicated up to $9.2 trillion of global economic losses resulting from not providing LMICs with diagnostics, therapeutics, and vaccines needed to manage COVID-19.[9]

The ability of pathogens to be transmitted across borders and the interconnectedness of global economies provided HICs with the greatest incentive to do something about vaccine maldistribution: their own self-interest.

From Eradicating Smallpox to Eliminating Other Infectious Diseases

When the WHO declared smallpox to be eliminated on May 8, 1980,[10] other infectious diseases were proposed as the next eradication targets, particularly childhood illnesses preventable through vaccination. Elimination of polio, diphtheria, pertussis, tetanus, and measles among children in LMICs became the catalyst for launching a new program, the *Expanded Programme on Immunization (EPI)*, by the World Health Assembly in 1974.[11]

EPI's first area of focus was the diphtheria-tetanus-pertussis vaccine (DTP3). The DTP3 vaccine (which protects against all three diseases in one comprehensive formulation) is administered in a five-shot series to children at two months, four months, six months, 15 to 18 months, and four to six years. DTP3 coverage by 12 months of age is now considered a global standard against which national childhood immunization programs are measured and compared.

The EPI's goal was to ensure that, by 1990, 80% of infants were current with the three DTP3 vaccines needed before their first birthday, and it was largely successful. By 1990 global DTP3 coverage reached 75%, with a 62% rate among low-income countries. However, DTP3 coverage plateaued or declined slightly over the following ten years due to a lack of sustained government commitment and health system inadequacies. As the twenty-first century dawned, it would take the efforts of new *public–private partnerships* to restore global DTP3 coverage.

Public–Private Partnerships: Gavi, CEPI, and COVAX

Twentieth-century global vaccination programs—to completely end smallpox and polio as human infectious diseases or to ensure children are inoculated against diphtheria, tetanus, and pertussis—were largely executed by United Nations-based agencies: the WHO and the United Nations Children's Fund (UNICEF).[12]

Ending world hunger, sustaining peace among nations, and controlling the global spread of infectious diseases are ambitious, complex, and expensive human endeavors requiring sustained effort, cooperation, and large amounts of money. Though UN-based agencies had some success with smallpox eradication and childhood immunization programs, the resurgence of vaccination-preventable diseases in the 1990s was a reminder that constant vigilance over these human tragedies was sorely needed.[13]

Familiar twentieth-century legacy groups important to disease eradication efforts—such as the WHO, UNICEF, and the World Bank—gained new purpose in confronting twenty-first century distribution challenges.

Their efforts were augmented by newly emerging twenty-first century groups: the Global Alliance for Vaccines and Immunization (Gavi) founded in 2000, the Coalition for Epidemic Preparedness and Innovation (CEPI) founded in 2017, and the COVID-19 Vaccines Access Facility (COVAX) founded in 2020. Though Gavi, CEPI, and COVAX are financially supported by many funding sources, one private group, the Bill and Melinda Gates Foundation, was a major financial contributor to global vaccination efforts.

The flagging of DTP3 vaccination rates was reversed by the intervention of Gavi in 2000, which not only donated funds and technical assistance to reinvigorate DTP3 coverage activities, but was also a major impetus to other vaccination programs to reduce or eliminate global polio, measles, meningitis, rotaviral-induced diarrhea, and yellow fever.[14]

Beginning in 2020, such entities became crucial in the effort to vaccinate the global population against SARS-CoV-2. Notably, these organizations have coordinated their efforts with one another, local governments, and other foundations to increase their reach and impact—creating an ecosystem of alliances aimed at both achieving global population vaccination and supporting vaccination rates in LMICs.

Bill and Melinda Gates Foundation

Private philanthropy has always been an important funding source for solving world problems, but the substantial contributions of the *Bill and Melinda Gates Foundation* are noteworthy. Indeed, the Gates Foundation and its initial pledge of $750 million in seed money to launch Gavi may be the bridge linking twentieth and twenty-first century global vaccination efforts.

Gavi, the Vaccine Alliance

Founded by the Bill and Melinda Gates Foundation and working in partnership with a plethora of government and private-sector foundations, *Gavi*'s mission is to "save children's lives and protect people's health by increasing equitable use of vaccines in lower-income countries."[15] This is measured in five ways: children immunized, future deaths prevented, under-five mortality rate reduced, future disability-adjusted life years averted because of vaccination, and sustained vaccine delivery within previously supported countries as Gavi involvement ends.

Gavi structures its work in five-year funding cycles, with its most recent cycle, Gavi 4.0, running from 2016 to 2020 and its fifth funding cycle, Gavi 5.0, beginning January 2021. Gavi has played a seminal role in SARS-CoV-2 vaccine distribution efforts through its partnership with COVAX.

Coalition for Epidemic Preparedness Innovation (CEPI)

The *Coalition for Epidemic Preparedness Innovation (CEPI)* is another important new actor in global vaccination activities. After attempts to develop, manufacture, and deliver vaccines against Zika, SARS, H1N1 influenza, and Ebola[16] stalled out, this new group formed in 2017 with representatives from eight countries (Australia, Belgium, Canada, Ethiopia, Germany, Japan, Norway, and the United Kingdom) and funding from the European Commission, Wellcome Trust, and the Gates Foundation.[17]

Since launching, CEPI has concentrated its efforts on developing vaccines against WHO Priority List pathogens, which are mostly antibiotic-resistant bacteria (acinetobacter, pseudomonas, and various enterobacteriaceae)[18] found in hospitals and nursing home settings. COVID-19 has shifted the group's priorities, and, in similar fashion to Gavi, CEPI is now playing an important role in global distribution of SARS-CoV-2 vaccine thorough its work with COVAX.

COVID-19 Vaccines Global Access Facility (COVAX)

HICs can purchase huge quantities of vaccine, negotiate their own favorable contracts with vaccine manufacturers, and mobilize considerable resources for impactful initiatives such as Operation Warp Speed, which was largely responsible for helping develop a safe and effective SARS-CoV-2 vaccine in eleven months (as discussed in Chapter 7: Vaccines).

LMICs, however, lack these capabilities and need vaccine donations and technical assistance to accomplish the key tasks of global vaccination. The SARS, H1N1 influenza, MERS, Ebola, and Zika epidemics clearly demonstrated the dangers of private vaccine manufacturers abandoning vaccine development projects and HICs monopolizing vaccine purchases through pre-license agreements. In the face of SARS-CoV-2, an alternative path for LMICs was needed.

As with previous global vaccination efforts, the WHO became the nexus for coordinating worldwide collaboration among HICs and LMICs. In April 2020, one month after COVID-19 was deemed a global pandemic, the WHO convened eight entities (CEPI, Gavi, Global Fund, Unitaid, Foundation for Innovative New Diagnostics, World Bank Group, Wellcome Trust, and the Gates Foundation) and launched a new joint effort called the *Access to COVID-19 Tools Accelerator (ACT-A)*.[19]

ACT-A was designed to expedite access for LMICs according to four pillars:

- Diagnostics pillar to increase access to COVID-19 testing materials
- Therapeutics pillar to increase access to medication, oxygen, and other therapeutic agents
- Vaccines pillar, also called the *COVID-19 Vaccines Global Access Facility (COVAX)*,[20] for vaccine distribution
- Health Systems Connector to assist with oversight, delivery, and logistics

By November 2020, ACT-A was able to make available to LMICs 120 million rapid SARS-CoV-2 tests; secure dexamethasone, oxygen, and monoclonal antibody treatment for close to three million people; and support development efforts for ten vaccine candidates.

Looking Back, Looking Ahead

Achieving global vaccination rates sufficient to stop the spread of COVID-19 requires tremendous resources and coordination to get vaccines from manufacturers into the arms of massive numbers of individuals. This goal is further complicated by resource inequities between HICs and LMICs. The twentieth century saw the advent of several successful mass-vaccination efforts, facilitated by international organizations, both public and private.

As the interconnectedness of global economies increased in the twenty-first century, new challenges arose in responding to EIDs. No longer could HICs focus primarily on their own populations; global vaccination coverage is required to stem the spread of pathogens across borders. New public–private partnerships arose to address this issue, and these organizations have been instrumental in the distribution of SARS-CoV-2 vaccines, especially in LMICs.

But as future COVID-19 variants and other EIDs appear, questions persist about whether these new organizations will be up to the tasks of global vaccination. The Gavi 5.0 funding cycle could be insufficient to address emerging needs in an uncertain landscape. ACT-A has been criticized for its "lack of ambition" and having its mission sullied by political squabbles, unkept promises, and its "heavy focus on technical solutions."[21] And reliance on massive funding from nonprofit organizations—even those as large and well-resourced as the Gates Foundation—places the fate of LMIC populations in the hands of individual philanthropists in HICs. How these questions are resolved will have profound effects on the possibility of eradicating EIDs—anywhere and everywhere—as the twenty-first century progresses.

FUTURE OUTINGS

To understand how the lessons of COVID-19 can be applied to future EIDs, ongoing excursions into the field must explore:

- The root causes and contributors to health inequities between HICs and LMICs
- Systemic solutions to the challenges of vaccination programs in LMICs
- Emerging barriers to the six tasks of global vaccination
- The ecosystem of alliances, both public and private, responding to EIDs

Student Research Questions

1. Health inequities between HICs and LMICs are rooted in many historical forces. What factors have contributed to the current state of

health inequity among nations? What roles have racism, colonialism, globalization, and political conditions played?

2. The challenges of vaccination in African countries were often successfully overcome in the efforts to curb the spread of diseases such as Ebola, meningitis, and yellow fever. How and why did these programs succeed? Where did these programs fall short? What lessons do these programs offer for future efforts to achieve vaccination targets in LMICs?

3. How have national healthcare policies and systems of various nations incentivized or de-incentivized pharmaceutical companies to research, develop, and manufacture vaccines? In what ways do the interests of private, for-profit companies intersect with government programs regarding public health? How have these factors affected global COVID-19 vaccination efforts, and what implications are there for future EID vaccination programs?

4. How have public, government-sponsored organizations interacted with private organizations in responding to COVID-19? What are the benefits and drawbacks of relying on funding from private philanthropists to respond to public health crises? What are the implications of public–private partnerships for global responses to future EIDs?

5. Explore the current state of Gavi, CEPI, and COVAX. Since their initial efforts to address COVID-19, what progress have they made toward their goals? How have their goals and strategy shifted? What is the current state of their funding and sphere of influence?

6. In what ways might the current ecosystem of public–private alliances need to evolve to confront the EID issues of the future?

NOTES

1. "WHO, UN set out steps to meet world COVID vaccination targets," October 7, 2021, https://www.who.int/news/item/07-10-2021-who-un-set-out-steps-to-meet -world-vaccination-targets, accessed October 27, 2021.
2. Mathieu, E., Ritchie, H., Ortiz-Ospina, E. et al. "A global database of COVID-19 vaccinations," *Nat Hum Behav*, 2021. "Share of people vaccinated against COVID-19, October 26, 2021," https://ourworldindata.org/covid-vaccinations, accessed October 27, 2021.
3. United Nations, Department of Economic and Social Affairs, Population Division (2019). World Population Prospects 2019, custom data acquired via website accessed March 14, 2022.

4. In the US in 2020, there were 19,301,292 children aged 0 to 4 years and 28,384,878 children 5 to 11 years. US Census data as of April 2020 indicated a total US population of 331,449,281 (sources: The Annie E. Casey Foundation, KIDS COUNT Data Center, https://datacenter.kidscount.org/data/tables/101-child-population-by-age -group#detailed/1/any/false/574/62,63/419 and US Census Bureau Quick Facts, https://www.census.gov/quickfacts/fact/faq/US/PST045221, both cites accessed March 14, 2022.

5. Liz Hamel, Lunna Lopes, Grace Sparks, Ashley Kirzinger, Audrey Kearney, Mellisha Stokes, Mollyann Brodie, "KFF COVID-19 Vaccine Monitor: October 2021," https://www.kff.org/coronavirus-covid-19/poll-findings/kff-covid-19-vac-cine-monitor-october-2021, accessed March 10, 2022.

6. Classifying countries by income status (gross national income or GNI per capita) proposed by the World Bank ("low-income countries" or LIC, "middle-income countries" or MIC, "low- or middle-income countries" or LMIC) is replacing "third world" or "developing country" as descriptive terms. For 2022, lower middle-income economies had GNI per capita between $1,046 and $4,095, with Nigeria as an example. High-income economies are those with GNI per capita of $12,696, with the United States as an example. (Source: The World Bank: World Bank Country and Lending Groups: World Bank Data Help Desk, https://datahelpdesk .worldbank.org/knowledgebase/articles/906519-world-bank-country-and-lending -groups, accessed March 14, 2022.

7. Mathieu, E., Ritchie, H., Ortiz-Ospina, E. et al. "A global database of COVID-19 vaccinations," *Nat Hum Behav*, 2021. "Share of people vaccinated against COVID-19, October 26, 2021," https://ourworldindata.org/covid-vaccinations, accessed March 12, 2022.

8. Official data collated by Our World in Data—Last updated May 17, 2022.

9. "The Economic Case for Global Vaccinations," © International Chamber of Commerce, January 2021, https://iccwbo.org/publication/the-economic-case-for -global-vaccinations/, accessed March 14, 2022.

10. "Commemorating smallpox eradication—a legacy of hope, for COVID-19 and other diseases," WHO news release, May 8, 2020, https://www.who.int/news/item /08-05-2020-commemorating-smallpox-eradication-a-legacy-of-hope-for-covid -19-and-other-diseases, accessed November 14, 2021.

11. WHO is the health agency within the United Nations, and the World Health Assembly is the main governing body of WHO. (Sources: "WHO and the WHA—an explainer," https://www.who.int/about/governance/world-health-assembly/seventy-third-world -health-assembly/the-who-and-the-wha-an-explainer, accessed March 14, 2022. See also Okwo-Bele JM, Cherian T. "The expanded programme on immunization: a lasting legacy of smallpox eradication," *Vaccine*, December 30, 2011, volume 29, supplement 4, pages D74–D79, doi: 10.1016/j.vaccine.2012.01.080. PMID: 22486980.)

12. Founded in 1946, the acronym "UNICEF" stands for "United Nations International Children's Emergency Fund" describing the agency's founding mission to assist the emergency needs of children in post-World War II Europe and China. Four years later its mission broadened to include the needs of women and children's in developing nations, and though its official name was shortened to "United Nations Children's Fund," the "UNICEF" acronym was retained (UNICEF Frequently Asked Questions, https://www.unicef.org/about-unicef/frequently-asked-ques-tions, accessed March 15, 2022).

13. Not all countries in the world are members of the United Nations. As of 2022 there were 193 member states recognized by the UN. According to Article 4 of the UN Charter "membership in the United Nations is open to all other peace-loving states" which does not characterize all countries (United Nations About Us, https://www .un.org/en/about-us, accessed March 15, 2022).

14. Ann Lindstrand, Thomas Cherian, Diana Chang-Blanc, Daniel Feikin, Katherine L O'Brien, "The World of Immunization: Achievements, Challenges, and Strategic Vision for the Next Decade," *The Journal of Infectious Diseases*, volume 224, Issue Supplement, 1 October 2021, Pages S452–S467, https://doi.org/10.1093/infdis/ jiab284.

15. Gavi, the Vaccine Alliance Annual Progress Report Year 5—2020.

16. Waning interest in developing vaccines for twenty-first century pathogens occurred for two reasons: (1) shortened time courses or low case numbers (for example, SARS caused 8,096 cases and 774 deaths within 26 countries in nine months from November 2002 to July 2003); and (2) inordinate research and development costs incurred by pharmaceutical manufacturers in developing new vaccines.

17. Nicole Lurie, Melanie Saville, Richard Hatchett, Jane Halton, "Developing Covid-19 vaccines at pandemic speed," *New England Journal of Medicine*, May 21, 2020, volume 382, number 21, pages 1969–1973.

18. WHO news release, "WHO publishes list of bacteria for which new antibiotics are urgently needed," February 27, 2017, https://www.who.int/news/item/27-02-2017 -who-publishes-list-of-bacteria-for-which-new-antibiotics-are-urgently-needed, accessed March 15, 2022.

19. WHO ACT-Accelerator FAQ, https://www.who.int/initiatives/act-accelerator/faq, accessed March 15, 2022.

20. The term "facility" as in making something easy to do or to obtain ("she's a natural musician and plays the piano with such facility") and not "facility" in terms of a structure or building ("the music department on campus has two facilities—a teaching venue and a performance venue").

21. "The ACT Accelerator: heading in the right direction?," *Lancet* editorial, volume 397, April 17, 2021, https://www.thelancet.com/journals/lancet/article/PIIS0140 -6736(21)00835-7/fulltext, accessed March 15, 2022.

10

Covidiocy

The Field: Why would otherwise rational people ingest dangerous chemicals or intentionally expose themselves to a deadly pathogen? Why would they deny the existence of a disease that was killing thousands of people in their own communities? While such behaviors may seem absurd, they were driven by the promulgation of confusion and misinformation that spread alongside the SARS-CoV-2 virus itself—often aided by the behaviors of public officials. COVID-19 can in this context be seen as a dual pandemic: a pathogen attacking the body and misinformation infecting the mind.

Field Sightings: "infodemic", misinformation, disinformation, Surgeon General's Advisory, "covidiocy", logical fallacies, bias, cultural values.

FIELD EXPEDITION: VECTOR IN CHIEF[1]

On Thursday, April 23, 2020, William Bryan, who was then acting head of the Department of Homeland Security's Science and Technology Directorate, attended the daily press briefing of the White House Coronavirus Task Force to review how sunlight, temperature, humidity, and bleach affected the SARS-CoV-2 virus. Bryan's department had been studying the issue because the precise way in which SARS-CoV-2 was transmitted was not entirely clear. Pre-empting the reporters waiting to ask Bryan questions, President Trump took the podium and made these remarks:

> Thank you very much. So I asked Bill a question that probably some of you are thinking of, if you're totally into that world, which I find to be very interesting. So, supposing we hit the body with a tremendous—whether it's

DOI: 10.1201/9781003310525-11

ultraviolet or just very powerful light—and I think you said that that hasn't been checked, but you're going to test it. And then I said, supposing you brought the light inside the body, which you can do either through the skin or in some other way, and I think you said you're going to test that, too. It sounds interesting ... And then I see the disinfectant, where it knocks it out in a minute. One minute. And is there a way we can do something like that, by injection inside or almost a cleaning. Because you see it gets in the lungs, and it does a tremendous number on the lungs. So it would be interesting to check that.[2]

The next day, Reckitt Benckiser Group, a British manufacturer of Lysol® and Dettol® cleaning products, issued a press release entitled "Improper Use of Disinfectants":

> Due to recent speculation and social media activity, RB (the makers of Lysol and Dettol) has been asked whether internal administration of disinfectants may be appropriate for investigation or use as a treatment for coronavirus (SARS-CoV-2). As a global leader in health and hygiene products, we must be clear that *under no circumstance* should our disinfectant products be administered into the human body (through injection, ingestion, or any other route). As with all products, our disinfectant and hygiene products should only be used as intended and in line with usage guidelines. Please read the label and safety information. We have a responsibility in providing consumers with access to accurate, up-to-date information as advised by leading public health experts. For this and other myth-busting facts, please visit Covid-19facts.com.[3]

This was not the first time Trump had recommended an unorthodox approach to treating COVID-19. From March 1 to April 30, 2020, Trump posted 11 tweets recommending use of the unproven medications hydroxychloroquine, chloroquine, and azithromycin as treatments for COVID-19; he also mentioned these medications 65 times during White House briefings. His tweets did not go unnoticed and had an "impression reach," or number of potential user views per one individual tweet, of over 78 million.[4]

Trump was a prolific user of Twitter, posting some 57,000 tweets on his @realDonaldTrump account[5], including over 11,000 tweets during the first years of his presidency from January 20, 2017 to November 2, 2019.[6] Though the majority of the President's 11,000 tweets were used to attack someone or something, those concerning COVID-19 scientific topics were often factually incorrect and misleading, contributing to the rampant spread of misinformation and disinformation spawned during the pandemic.

People who acted on this misinformation and disinformation did so at their own peril:

- "Covid parties" were organized to intentionally expose healthy people to SARS-CoV-2 as either a dare or to gain immunity. A 30-year-old San Antonio, Texas man who contracted COVID-19 during one such party told his hospital nurse "I think I made a mistake. I thought this was a hoax but it's not…" and then died.[7]
- Upon learning hydroxychloroquine might prevent COVID-19, a married Arizona couple in their sixties who raised koi fish self-medicated themselves by drinking fish parasite remover containing chloroquine phosphate, which sounds like "hydroxychloroquine" but is far more potent. After drinking the solution, both were immediately sickened: the wife became critically ill and recovered; the husband had a cardiac arrest and died.[8]
- Herman Cain, a 2012 presidential contender and Godfather's Pizza CEO, attended President Trump's June 20, 2020 indoor campaign rally in Tulsa, Oklahoma, where he was photographed without a facemask among throngs of people. Nine days later he tested positive for SARS-CoV-2, perhaps contracted at the Tulsa event, was hospitalized, and died a month later on July 30, 2020.[9]

Like Herman Cain, the President was ambivalent about facemask wearing, mocking some for doing so on his Twitter account and advising others to wear masks at their own choosing. The President himself was not seen publicly wearing a facemask for the first seven months of the pandemic until Saturday, July 11, 2020 when he visited injured service members at the Walter Reed National Military Medical Center.

LESSON LEARNED: VECTOR IN CHIEF

The presidential "bully pulpit" is a powerful source of influence in American culture, and its power was displayed in new ways during the

"Twitter presidency" of Donald Trump. His unscripted and sometimes cringeworthy pronouncements during frequent appearances at the White House Coronavirus Task Force press briefings were replete with factual and misleading errors that had to be retracted by White House staff and, in the case of disinfectant use as a COVID-19 cure, publicly corrected by a manufacturer's press release.

As a never-before-seen virus swept the world, so did a glut of theories, speculation, opinions, and information—both scientifically sound and fallacious. The dangers posed by confusion and falsehoods would prove as great a threat to public health as the SARS-CoV-2 virus itself.

CONCEPTS COVERED IN THIS CHAPTER

- The COVID-19 infodemic
- Covidiocy
- Irrational behaviors
- Underlying causes of irrational behaviors

The COVID-19 Infodemic

Though President Trump's statements cannot be said to be directly responsible for harms caused by COVID-19, their sheer prolixity on Twitter and other social media platforms contributed to a growing climate of pandemic-related misinformation and disinformation. This growth in COVID-19 information is well-described by the portmanteau[10] *"infodemic,"* from "information" and "epidemic."[11]

The World Health Organization (WHO) defines infodemic as

> too much information including false or misleading information in digital and physical environments during a disease outbreak. It causes confusion and risk-taking behaviors that can harm health. It also leads to mistrust in health authorities and undermines the public health response.[12]

The WHO describes three aspects of the infodemic in its definition: the overabundance of information, the misuse of information, and the harmful consequences resulting from this misinformation.

The Overabundance of Information

The overabundance of information should not be surprising, since SARS-CoV-2 was a completely new virus that caused a global pandemic with lethal results. An example of this information overabundance is a PubMed search for references about COVID-19 related treatments over the first year of the pandemic (from February 1, 2020 to March 30, 2021) conducted by the NIH COVID-19 Treatment Guidelines Panel. This group found 57,241 published references to COVID-19 treatments[13] during this time period: an astounding number of scientific articles.

The Misuse of Information

Misuse of information may be categorized in two ways: misinformation and disinformation. *Misinformation* is health information that is "false, inaccurate, or misleading according to the best available evidence at the time," and *disinformation* is misinformation "spread intentionally to serve a malicious purpose."[14] The dissemination of misinformation and disinformation has been supercharged by three social media companies: Google, Twitter, and Meta. Meta owns Facebook, WhatsApp, and Instagram, and these applications have a global reach encompassing a vast portion of the planet's population: Facebook with 3 billion estimated users, WhatsApp at 2 billion, and Instagram 1.4 billion.[15]

The Harmful Consequences of Misinformation

By July 2021 the harmful consequences of misinformation pushed the US Surgeon General to issue a *Surgeon General's Advisory* entitled "Confronting Health Misinformation: The U.S. Surgeon General's Advisory on Building a Healthy Information Environment."[16]

Surgeons General are appointed by the President to represent the federal government's views on public health, and since the first Surgeon General was appointed in 1871 to lead the Marine Hospital Service, Surgeons General have retained leadership of the Commissioned Corps of the Public Health Service. Surgeons General began issuing scientific Advisories about public health issues in the 1960s, with the first statement coming from Surgeon General Leroy E. Burney in 1964 entitled "Smoking and Health: Report of the Advisory Committee to the Surgeon General" about the causal link between cigarette smoking and a number of disease states and poor health outcomes. Since that initial 1964 report, Surgeons General have selectively issued official Advisories only when public health matters have risen to a level of serious national concern. This is best exemplified by the October 1986 release of "The Surgeon General's Report on Acquired Immune Deficiency Syndrome" by Surgeon General C. Everett Koop during another period in American history dominated by health misinformation and prejudicial reasoning: the AIDS crisis of the 1980s.

Like Surgeons General Burney and Koop, the Biden administration's Surgeon General Vivek H. Murthy was compelled to issue his July 2021 Advisory about COVID-19 related misinformation because people were increasingly taking untested medications or remaining unvaccinated at a time when safe and effective vaccines were available. His report opens with this admonishment:

> I am urging all Americans to help slow the spread of health misinformation during the COVID-19 pandemic and beyond. Health misinformation is a serious threat to public health. It can cause confusion, sow mistrust, harm people's health, and undermine public health efforts. Limiting the spread of health misinformation is a moral and civic imperative that will require a whole-of-society effort.[17]

Vivek's Health Advisory provides examples of actions which communities, health organizations, journalists, and other segments of society can take to confront health misinformation. It is indeed sobering to consider that the dangers of following health misinformation is now cast in the same light as the risks posed by smoking cigarettes, whose packaging is required to prominently display warnings such as "WARNING: Smoking causes head and neck cancer," often accompanied by a stark image of this serious problem.

Covidiocy

The COVID-19 infodemic is characterized by a gargantuan amount of health information, some of which is true and helpful, and some which is misleading and harmful. This information is aimed at an audience of worried, confused, and anxious people who need to process, synthesize, and use this information to make personal and family health decisions, which in many cases can lead to lethal outcomes. This audience has its own set of personal biases and intellectual resources, through which this overabundance of information is filtered. Given the sheer volume of information and opinion to be processed, people must often rely on the advice and examples of others they trust to help them make decisions about what to believe and do.

As some individuals in positions of public notoriety decided to seed the airwaves and social media spheres with erroneous advice—which was then picked up, spread, and sometimes followed by health consumers—it is not surprising to see numerous examples of irrational behaviors like "Covid parties" or drinking fish parasite remover containing chloroquine. These odd behaviors are well-described by the term "*covidiocy*," another portmanteau combining "covid" and "idiocy."

Covidiocy is not a term of indictment but a way to describe behaviors observed during the pandemic that were not in the self-interest of the people engaging in such actions. Observable behaviors, no matter how irrational they seem from an objective scientific viewpoint, have underlying causes, which should be explored without judgment. The life-altering decisions people made for themselves during the pandemic revealed how humans assess risk, evaluate institutional credibility, and decipher evolving streams of information during the sustained stress and uncertainty of an ongoing infectious disease emergency.

Irrational Behaviors and Beliefs Observed during the Pandemic

The pandemic was replete with "covidiotic" examples of irrational behaviors. These included denying the existence of the virus entirely, willfully engaging in extreme behaviors like toilet seat licking or spraying others with disinfectant, refusing quarantine and mask mandates, and using dangerous or discredited treatments for COVID-19 like hydroxychloroquine or ivermectin. Some of these unconventional therapeutic approaches came from a minority of physicians who themselves advanced seemingly rational and fact-based arguments, giving their viewpoints the patina of authority, when in reality there was no scientific basis for their claims; by advancing these arguments, these healthcare professionals practiced a kind of charlatanism generally not seen in modern medicine.[18]

Mythologies and conspiracy theories were borne out of this pool of misinformation, perpetuating treatment decisions made by vulnerable people. For instance, some people continue to incorrectly believe COVID-19 vaccines induce miscarriages or cause infertility, based on social media posts claiming that mRNA vaccine-developed antibodies against SARS-CoV-2 spike protein could theoretically cross-react with human placental protein syncytin-1 because these two proteins share an insignificant number of amino acid sequences. In fact, COVID-19 vaccinated pregnant women are no more likely to miscarry their pregnancies than unvaccinated pregnant women, and studies have shown that pregnant women are more likely to become severely ill or die from COVID-19 than nonpregnant women and should therefore be vaccinated.[19] This is one example of many conspiracy theories and factually-unsound beliefs that have run rampant since the beginning of the pandemic; others include that the virus was spread by 5G radiofrequencies, that Bill Gates caused the COVID-19 pandemic in order to implant tracking microchips into people through vaccines, that governments are reporting falsely high death rates, and that COVID-19 was a hoax perpetrated by the "deep state" or a global cabal to stifle Americans' freedoms.

While many conspiracy theories made extreme claims that struck most people as unlikely, they contributed to a general undermining of reliable facts. However, less outlandish beliefs were common, including that the virus was unlikely to cause serious illness in otherwise healthy persons, that contracting the virus was not likely, and that vaccination was unnecessary or potentially dangerous.

The results of these irrational behaviors and beliefs was both tragic and unsurprising: increased illness and death.

Underlying Causes and Driving Forces for Irrational Behavior

Why did so many people behave in ways that were detrimental to their own health—and that of others—in the midst of a global pandemic? What factors contributed to decision-making that seems to defy reason during a time in which sound reasoning is of paramount importance? The seeds of misinformation and disinformation often found fertile ground due to three factors—logical fallacies, bias, and cultural values—all of which are heightened in an environment of fear and uncertainty.

Logical Fallacies

The COVID-19 pandemic was experienced as a once-in-a-lifetime event with life-altering, and in some cases lethal, consequences. This was especially the case before COVID-19 vaccines became available, when many people lived in mortal fear. Fear and uncertainty undermine critical thinking, impeding the ability to make rational decisions and causing some to seize on fallacious arguments which may sound convincing or appeal to heightened emotions. *Logical fallacies*[20] may stem from logical thinking that is based in faulty premises or from rationalization that leads to conclusions not actually supported by logical reasoning. One common logical fallacy related to COVID-19 vaccines has been the "appeal to ignorance" fallacy, in which a lack of evidence for one outcome is used as proof of evidence for a different outcome. The thinking in this fallacy states that because the long-term side effects of COVID-19 vaccines are not known, the vaccines must not be safe. This irrational conclusion elevates what is *not* known and ignores the evidence supporting the safety of COVID-19 vaccines. For people who lack an understanding of how vaccines function and are feeling fearful, this line of reasoning may sound logical, even though it is not.

Bias

All people unconsciously harbor prejudicial beliefs, which may be xenophobic, homophobic, racist, sexist, ageist, and misogynistic. This constellation of human traits influences an individual's decision-making in all circumstances and is often heightened in times of uncertainty. A prime example of how *bias* informed beliefs and behaviors during the early days of the pandemic is the belief—especially prominent in America—that the SARS-CoV-2 virus was created and leaked, perhaps intentionally, from a laboratory in China. Attitudes of xenophobia against people from China, combined with a history of tension between the Chinese and American governments, allowed this bias to color some people's beliefs about the origins of the virus, with some prominent politicians, including Donald Trump, referring to SARS-CoV-2 as "the Chinese virus." The results of bias are not abstract; the Stop AAPI Hate coalition received reports of 10,905 hate incidents against Asian American and Pacific Islander persons between March 19, 2020 and December 21, 2021.

Cultural Values

All humans live within communities with their own unique *cultural values*, political structures, and governmental organizations, and a public health crisis is experienced within the framework of these values, beliefs, and structures. American culture was founded upon a distrust of centralized authority, originating in the Declaration of Independence and codified in the Constitution. Americans place a high value on their "unalienable rights to life, liberty, and the pursuit of happiness,"[21] and during the COVID-19 pandemic the personal expression of individual freedom felt, to many, at odds with the measures needed to protect others through quarantines and vaccination programs.

Vaccination rates may reflect how the cultural values of a country affect public health initiatives. As of March 1, 2022, 76% of Americans had received at least one dose of vaccine; 65% of those had completed a full vaccination protocol. In contrast, Denmark had a vaccination rate of 83%, with 82% fully vaccinated. Denmark's culture prizes social trust and individual investment in the good of the whole community. Danes pay some of the highest taxes in the world, financing an extensive network of social welfare programs, and consistently evince high degrees of trust in government authority. While not definitively causative, a correlation can be seen between cultural values and the portion of the population that chose to be vaccinated in these two countries.

Looking Back, Looking Ahead

A potentially deadly virus, copious amounts of changing information, a global instantaneous communication infrastructure, frightened vulnerable people, and actors with their own agendas produced catastrophic results during the pandemic, especially in the United States, which reached the grim milestone of one million COVID-19 related deaths on May 17, 2022—the highest total in the world.

The COVID-19 infodemic reveals how susceptible people can become to misinformation and unscientific judgment in the midst of uncertainty and fear—and covidiocy has been a natural result. Future emerging infectious diseases (EIDs) are likely to arise amidst a similar landscape—perhaps more so as technology enables ever more rapid transmission of vast quantities of information, opinions, and distortions. In the same way that certain populations are more susceptible to infection and severe illness from SARS-CoV-2, individuals may be more or less likely to be swayed into irrational conclusions based on the biases they hold and cultural contexts in which they live. When the resultant behaviors produce harm, it is imperative to understand their underlying causes in order to mitigate the damage.

Public officials, researchers, and healthcare professionals of the future must be prepared to fight not just new viruses, but accompanying misinformation that spreads as quickly as disease—and with similarly deadly results.

FUTURE OUTINGS

To understand how the lessons of COVID-19 can be applied to future EIDs, ongoing excursions into the field must explore:

- The avenues by which misinformation is spread and tools for curbing it
- The role of human psychology and media systems in inciting and promulgating an infodemic
- The role of educational systems in preparing populations to effectively ingest and interpret scientific information

Student Research Questions

1. What people and institutions were the primary sources of misinformation during the pandemic? Through what channels was this misinformation released and spread?
2. How did major American media outlets report on factually inaccurate or questionable statements made by persons in positions of authority? What level of fact-checking or context accompanied the reporting on such statements?
3. What harmful behaviors did people engage in at different phases of the COVID-19 pandemic? What was the source of the information or opinion that sparked these behaviors? What were the results of these behaviors on the individuals that engaged in them and on the spread of disease?

4. How did regulatory agencies and scientific organizations respond to misinformation that appeared in news outlets and on social media? How were those responses received by the public?

5. What were the logical fallacies behind the most common conspiracy theories and misjudgments about the COVID-19 pandemic?

NOTES

1. A vector is "an organism (such as an insect) that transmits a pathogen from one organism or source to another" ("vector." Merriam-Webster.com. 2011. https://www.merriam-webster.com, accessed May 26, 2022). "Vector in Chief" is a pun based on "Commander in Chief," one of the principal duties of an American president. Fintan O'Toole used this moniker to describe the personal actions of President Donald Trump during the COVID-19 pandemic in his May 14, 2020 New York Review of Books essay "Vector in Chief."

2. White House Coronavirus Task Force press briefing transcript, April 23, 2020.

3. Reckitt Benckiser Group press release, "Improper use of Disinfectants," April 24, 2020, www.reckitt.com/newsroom.

4. This observational study "investigated the relationship between Donald J Trump's advocacy for unproven treatments and the media landscape of COVID-19 treatments" and found that "Donald J Trump's first tweet on March 21, 2020 advocating for the use of hydroxychloroquine and azithromycin was one of his most popular tweets. With 385,700 likes and 103,200 retweets at the time of this study, the tweet had a potential estimated impression reach (i.e., the number of potential user views on the individual tweet) of 78,800,580" (Niburski K, Niburski,O. "Impact of Trump's Promotion of Unproven COVID-19 Treatments and Subsequent Internet Trends: Observational Study," *J Med Internet Res.*, 2020, volume 22, number 11, page e20044. Published November 20, 2020. doi:10.2196/20044).

5. @realDonaldTrump was first used by Trump on May 4, 2009 to announce his appearance on Late Night with David Letterman and continued in use until Twitter permanently suspended the @realDonaldTrump account on January 8, 2021 as a result of tweets about the 2020 election. Twitter assessed two tweets posted by Trump on January 8, 2021 finding they violated their "Glorification of Violence policy which aims to prevent the glorification of violence that could inspire others to replicate violent acts and determined that they were highly likely to encourage and inspire people to replicate the criminal acts that took place at the US Capitol on January 6, 2021."

6. Shear M, Haberman M, Confessore N, Yourish K, Buchanan L, Collins K. "How Trump Reshaped the Presidency in Over 11,000 Tweets," *New York Times*, November 2, 2019.

7. Pietsch B. "Man, 30, dies after attending a 'Covid Party', Texas hospital says." *New York Times*. July 12, 2020.

8. Stokkermans TJ, Goyal A, Bansal P, et al. "Chloroquine and hydroxychloroquine toxicity" (updated August 25, 2021). *StatPearls* (internet). Treasure Island, Florida. StatPearls Publishing, January 2021. www.ncbi.nlm.nih.gov/books/NBK537086, accessed May 26, 2022.

9. Ortiz A, Seelye K. "Herman Cain, former C.E.O. and presidential candidate, dies at 74." *New York Times*. July 30, 2020.

10. A portmanteau combines two unrelated words to create a new word which derives its meaning from the blended word roots. For instance, "motel" is a portmanteau of "motor" and "hotel" describing roadside lodging, "podcast" blends "iPod" and "broadcast" for a content-rich program accessible on a mobile device.

11. "Infodemic" was first used in 2003 by Washington Post columnist David Rothkopf describing fallout from SARS. "SARS is the story of not one epidemic but two, and the second epidemic, the one that has largely escaped the headlines, has implications that are far greater than the disease itself. That is because it is not the viral epidemic but rather an 'information epidemic' that has transformed SARS, or severe acute respiratory syndrome, from a bungled Chinese regional health crisis into a global economic and social debacle…the information epidemic—or 'infodemic'— has made the public health crisis harder to control and contain." "When the Buzz Bites Back," David J. Rothkopf, *Washington Post*, May 11, 2003.

12. WHO, "Infodemic," https://www.who.int/health-topics/infodemic#tab=tab_1.

13. Kuriakose S, Singh K, Pau AK, et al. "Developing treatment guidelines during a pandemic health crisis: lessons learned from COVID-19," *Ann Intern Med.*, 2021, [Epub ahead of print]. Doi:10.7326/M21-1647. PubMed has over 33 million citations and abstracts from biomedical literature and is maintained by NIH.

14. Office of the Surgeon General (OSG). Confronting Health Misinformation: "The U.S. Surgeon General's Advisory on Building a Healthy Information Environment [Internet]. Washington (DC): US Department of Health and Human Services; 2021. Available from: https://www.ncbi.nlm.nih.gov/books/NBK572169.

15. Gisondi M, Barber R, Faust J, Raja A, Strehlow M, Westafer L, Gottlieb M. "A Deadly Infodemic: Social Media and the Power of COVID-19 Misinformation," *J Med Internet Res*, 2022; volume 24, number 2, page e35552. https://www.jmir.org /2022/2/e35552, doi: 10.2196/35552.

16. See Note 14.

17. See Note 14.

18. One such group, America's Frontline Doctors, published a *White Paper on Hydroxychloroquine* to "draw the reader's attention to the indisputable safety of hydroxychloroquine (HCQ), an analog of the same quinine found in tree barks that George Washington used to protect his troops" and also promulgated the February 10, 2021 brief *Expanding Use of Ivermectin as Early Treatment for COVID* stating "uneven COVID-19 vaccine distribution and administration, and justifiable vaccine skepticism due to its unprecedented speed, provides additional impetus for employing safe repurposed anti-infective therapies that have been time-tested and are widely available." Positions presented in these papers supporting the use of hydroxychloroquine and ivermectin for COVID-19 are based on fallacious premises. As previously discussed, hydroxychloroquine and ivermectin are both FDA approved, but they are approved for therapeutic uses other than for COVID-19: hydroxychloroquine is FDA approved for use in treating malaria and ivermectin for use as an antiparasitic agent.

19. See two November 19, 2021 articles published in volume 70 of *CDC's Morbidity and Mortality Weekly Report*: "COVID-19 associated deaths after SARS-CoV-2 infection during pregnancy—Mississippi, March 1, 2020–October 6, 2021" by Kasehagen L et al. and "Risk for stillbirth among women with and without COVID-19 at delivery hospitalization—United States, March 2020–September 2021" by DeSisto C et al.

20. "Logical fallacy is the reasoning that is evaluated as logically incorrect and that undermines the logical validity of the argument and permits its recognition as unsound. Logical fallacy can occur as accidental or can be deliberately used as an instrument of manipulation," from "Logical Fallacies," Domina Petric, MD, *The Knot Theory of Mind*, February 15, 2020, doi: 10.13140/RG.2.2.24781.18401/1.

21. Paraphrased from opening of the Declaration of Independence.

Conclusion: Our Pandemic Future

THE WHITE HOUSE CORRESPONDENTS' DINNER

The White House Correspondents' Dinner is a 100-year-old Washington, DC tradition bringing together journalists, politicians, celebrities, and the President for an evening of light-hearted fun. Playfully dubbed "the nerd prom", hundreds vie for a ticket to the evening's festivities, emceed by a noteworthy comedian and featuring self-deprecating humor by the President.

But the May 7, 2022 affair turned out to be, in the words of comedian Trevor Noah, the evening's emcee, "the nation's most distinguished superspreader event."[1] About 2,600 guests were packed into the Washington Hilton ballroom, and by the following Wednesday several attendees, including Secretary of State Anthony Blinken, had tested positive for COVID-19.

Notably absent from the festivities that evening was Dr. Anthony Fauci, the 81-year-old COVID-19 medical adviser to the President and long-serving Director of the National Institute of Allergy and Infectious Diseases. His choice to abstain from the event was, in his words, based on "individual assessment of my personal risk."[2]

A few weeks earlier, some 700 guests attended another large DC celebrity event, the Annual Gridiron Dinner, which sickened 72 people, including Speaker of the House Nancy Pelosi, Attorney General Merrick Garland, and the President's sister, Valerie Biden Owens.

Slightly more than two years since the WHO declared COVID-19 a global pandemic, with infection still rampant even as vaccines and therapeutics were reducing illness and death, many of the highest-ranking officials in the US would seem to have deemed their personal risk of contracting COVID-19 low enough to return to the kind of carefree gatherings they enjoyed before SARS-CoV-2 swept the world. They were not alone: the American public, tired of mask mandates and social constraints, yearned for a return to normalcy and resumption of pre-pandemic activities.

PERSONAL RISK ASSESSMENT

The fact that many attendees at the Correspondents' Dinner and Gridiron Dinner in spring 2022 fell ill with COVID-19—and subsequently exposed an unknown number of others to the virus—is not surprising. The Omicron strain had demonstrated its superior transmissibility out of all previous SARS-CoV-2 variants, peaking in January 2022 and continuing through May 2022. Infections were commonplace even among vaccinated and boosted people who had managed to evade contracting COVID-19 in previous months.

Though Omicron infections had plummeted from January highs, certain states began to see increasing cases by May, but lagging indicators of hospitalizations and deaths increased only slightly if at all, demonstrating the effectiveness of COVID-19 vaccination efforts. By May 9, 2022, 83% of Americans eligible to be vaccinated had received at least one dose of vaccine.[3]

Dr. Fauci's decision to skip the White House Correspondents' Dinner due to his personal risk assessment seemed a prudent course of action, one that others would be wise to emulate, especially older people and those with chronic medical or immunocompromising conditions.

A personal risk assessment calls for carefully evaluating one's individual level of vulnerability, determining the likelihood of viral transmission for any gathering (number of people, inside location, proximity to others), and ensuring one's currency with vaccinations and boosters. It also requires vigilance for signs or symptoms of SARS-CoV-2 infection and the ability to quickly access antiviral therapeutic agents like Paxlovid™ or remdesivir to further reduce disease progression, hospitalization, or death should one become infected.

Of course, the reliability of any personal risk assessment depends on many factors: understanding one's true vulnerability to contracting an illness, correctly interpreting the most current information about infection rates and virus transmissibility, and the ability to access vaccines and pay for health care, among others. It is a calculation made with the inevitably limited information available, and it exists within the context of social pressures, competing needs for safety and social contact, and many unknowable variables.

Personal risk assessment is focused on an individual's understanding of their own likelihood of being exposed to and contracting a disease. Each person's tolerance for risk varies, and every individual choice reverberates throughout a community. Alongside one's own vulnerability there exists the vulnerability of others: family members, friends, strangers one encounters—and all the people *those* people will encounter. Personal risk assessment must answer the question of what we need for ourselves; social responsibility must examine the question of what we owe to others.

Since the start of the COVID-19 pandemic, individuals have balanced their personal risk assessment with their sense of social responsibility to varying degrees. But there is a larger factor at play in evaluating risk at both the personal and societal level: the next pandemic-level emerging infectious disease (EID).

NEW THREATS ON THE HORIZON

As the COVID-19 pandemic continues, new EID threats are already emerging. Even before COVID-19 arose, government officials had been honing their pandemic preparedness plans for what was dubbed "Disease X", an undefined viral illness that seemed likely to eventually emerge. Similar to the "2019 novel coronavirus" we now know as COVID-19, Disease X will likely be caused by an unidentified RNA virus from a zoonotic source with potential to inflict devastation on human populations. This dire prediction is increasingly possible given the encroachment of humans on animal habitats, climate change making certain environments more hospitable to arthropods and other insect disease vectors, and continuing global interconnectedness.

The 2012 MERS epidemic did not develop into a global pandemic in large part due to its virulence: its ability to replicate was so strong that it killed its hosts before they could spread disease globally. SARS-CoV-2 strikes a more advantageous balance for the virus, as enough infected people survive illness to spread the virus throughout populations. Its mutability has allowed ever-more transmissible variants to emerge. How much more effective will the next novel coronavirus be at striking this delicate balance, enabling rampant infection while sparing enough lives to continue its spread?

New Variants and COVID-24

In May 2022, two more Omicron subvariants (BA.4 and BA.5) were causing dramatic rises in COVID-19 cases in South Africa, a country with only 36% of its population vaccinated.[4] As noted in Chapter 9: Global Concerns, nearly all African countries have suboptimal vaccination rates and will continue to be a source of new SARS-CoV-2 variants. Some researchers even anticipate the emergence of another novel coronavirus within two years: "2024 novel coronavirus," or COVID-24.

H5N1 Avian Flu

May 2022 also saw a global outbreak of a new strain of H5N1 Avian flu among commercial poultry farms and wild birds, accompanied by a small number of bird-to-human transmission. Though few in number, epidemiologists are leery of such outbreaks and concerned about a repeated bout of H1N1 influenza that engulfed the world in 1918, killing 50 million people.

Monkeypox

In mid-May 2022, just as the world began to cautiously enter a COVID-19 endemic phase, where SARS-CoV-2 cases continued to appear but did not seem to cause the kind of serious infection, hospitalization, or death seen during previous waves of illness, a new scourge appeared: monkeypox.

Monkeypox is in the same virus family as smallpox, both orthopox viruses causing disfiguring skin lesions and death. Monkeypox was discovered in 1958 after two outbreaks of pox-like disease occurred in research monkeys (thus, the name "monkeypox"), with the first human cases occurring in 1970 in the Democratic Republic of Congo. Since then, monkeypox has been largely confined to African countries, but in the true fashion of EIDs, monkeypox developed recently in the non-endemic locations of Australia, Europe, and the United States.

The first recent US case of monkeypox occurred in Massachusetts on May 18, 2022, when the state's Department of Public Health reported a positive test result in a patient returning from Canada with a characteristic skin rash.

Since then, there have been 257 confirmed and 120 suspected cases worldwide and one death in Nigeria, with 12 US cases and no deaths as of May 26, 2022.

Monkeypox is transmitted by direct contact with bodily fluids and sores, indirect contact with "fomites" (clothing, towels, or bedding) contaminated with virus, and large respiratory droplets after close prolonged face-to-face contact. Most US cases occurred among men who have sex with men.

Unlike SARS-CoV-2, monkeypox is not yet considered to be a "potential pandemic pathogen" given its transmissibility, but the alarm caused by its appearance is a reminder that the world will need to be on its guard against emerging infectious diseases from known pathogens in new geographic locations, like monkeypox, as well as completely new pathogens like SARS-CoV-2.

WHAT'S OLD IS NEW AGAIN: RE-EMERGING DISEASES

In the twentieth century, tremendous advances in vaccine development and vaccination programs transformed the landscape of infectious diseases, reducing illness throughout the world. These efforts were especially effective in America and Europe, where many diseases were eliminated.

When a disease stops circulating in a region, it is considered *eliminated* in that region. If a particular disease is eliminated worldwide, it is considered *eradicated*. To date, only one infectious disease that affects humans has been globally eradicated: smallpox. Numerous other diseases have been eliminated in the US—most notably, several childhood diseases that had commonly killed or disabled millions of babies and children for generations—with only infrequent instances of limited outbreaks (see Table 11.1).

TABLE 11.1

List of Common Infectious Diseases Eliminated through
Vaccination Programs in the US during the Twentieth Century

Disease	First vaccine developed	Eliminated in the US
Polio	1955	1994
Measles	1963	2000
Rubella	1969	2004
Diphtheria	1925	2004

Unfortunately, many of the gains made in fighting these diseases began to ebb in the twenty-first century. An eliminated disease can only remain in check with ongoing vaccination among most of a population. The rise of "anti-vax" movements has led to a steady decline in the rate at which parents vaccinate their children, the result of which has been growing outbreaks of infectious diseases that had been considered relics of the past. Examples include:

- Measles: 1,282 cases of measles were reported in the US in 2019, the largest outbreak since 1992; 89% of those infected were unvaccinated or did not have documented vaccination.
- Mumps: while not declared eliminated, incidence of the disease declined by 99% after the widespread use of a vaccine in 1968, dropping from 152,209 cases in 1968 to just 231 in 2003. However, cases soared to 6,584 in 2006, and outbreaks have continued.
- Whooping cough: the first vaccine against this disease was developed in 1914 and was subsequently combined with vaccines against diphtheria and tetanus into the DTP vaccine in 1948. Widespread administration of DTP caused a dramatic drop in cases during the second half of the twentieth century, from 120,718 in 1950 to 7,298 in 1999. However, infection rates rose in the twenty-first century, with a notable spike of 48,277 cases in 2012.

In addition to new EIDs, public health is now being threatened by the re-emergence of infectious diseases for which effective prevention has long been the norm in the US and Europe. Compounding this threat is the fact that many of these illnesses have never been effectively contained in low-income countries due to insufficient funds and infrastructure for widespread vaccination programs.

OUR PANDEMIC FUTURE

SARS-CoV-2 is now an enduring part of the human experience: it may be temporarily quelled in one area while resurging in another; new variants resistant to current vaccines and treatments may arise; or it could become a relatively minor threat, like the seasonal flu, that sickens people every year but can be controlled with annual vaccines. Even if a threshold of herd immunity to SARS-CoV-2 is reached and incidences of illness become a distant memory, the fundamental shifts the virus has caused in our collective perception of public health are here to stay—as is the wariness about new pandemic-potential EIDs.

However the future unfolds, the COVID-19 pandemic has irrevocably changed the landscape in which we live, as will the next EID to arise— and the next after that. It is the task of all those in the health sciences and public policy realms to step into the continuing story of EIDs in the twenty-first century and continue mapping the field of our future.

NOTES

1. Chris Cameron, "Virus cases grow after White House Correspondents' Dinner," *New York Times*, updated May 9, 2022.
2. See Note 1.
3. Only persons five years of age and older were eligible for COVID-19 vaccination in May 2022. Vaccination rate from CDC COVID Data Tracker, May 9, 2022.
4. Mathieu, E., Ritchie, H., Ortiz-Ospina, E. et al. A global database of COVID-19 vaccinations. *Nat Hum Behav* (2021).

Glossary

Access to COVID-19 Tools Accelerator (ACT-A): joint effort among CEPI, Gavi, Global Fund, Unitaid, Foundation for Innovative New Diagnostics, World Bank Group, Wellcome Trust, and the Gates Foundation to expedite LMIC access to diagnostic tools, therapeutics, and vaccines.

acute respiratory distress syndrome (ARDS): life-threatening illness caused by pneumonia, severe burns, or trauma producing a severe drop in blood oxygenation and "stiff lungs" upon mechanical ventilation.

adaptive immune system: second line of immunologic defense against foreign invaders comprising antigen-presenting cells, T-cells, and B-cells.

adenine: purine nucleotide found in DNA and RNA molecules that pairs with thymine.

ADME (absorption, distribution, metabolism, and excretion): how therapeutic agents are processed from ingestion through elimination, described by pharmacokinetics.

Administrative Procedures Act (APA): 1946 federal law enacted during the Roosevelt administration expanding powers to regulatory agencies.

Advisory Committee for Immunization Practices (ACIP): CDC advisory group which reviews scientific evidence for vaccine candidates and makes recommendations to the CDC Director.

alveolus-capillary interface: thin tissue barrier through which oxygen and carbon dioxide gases are exchanged.

amino acids: molecules which combine to form proteins.

angiotensin converting enzyme 2 (ACE2): cellular receptor used by SARS and SARS-CoV-2 viruses to enter host cells, ACE2 is part of the renin-angiotensin-aldosterone physiologic system responsible for managing blood pressure and fluid balance.

animal reservoir: animals having the ability to maintain pathogens in a viable state without becoming infected by these pathogens, a source of zoonotic illness in humans.

antibody: blood protein made by plasma cells in response to an antigen.

antigen: any substance which elicits an immune response.

antigen rapid diagnostic test (Ag-RDT or RDT): laboratory or home-based test using lateral flow detection to determine presence of antigenic protein fragments within a specimen sample.

astrobiology: the study of the origin, evolution, distribution, and future of life in the universe.

asymptomatic: having no detectable symptoms.

Baltimore system of classification: a system for classifying viruses based on viral genome genetic material (RNA or DNA), conformation of viral genome genetic material (single or double-stranded), and orientation of the coding strand if the genome is single-stranded RNA (positive sense or negative sense), conceived by David Baltimore.

basic reproduction number (R0 or R-naught): estimated number of healthy people who would be made ill by one infected person. The zero or naught refers to the first generation of infection cases occurring at the start of a developing epidemic assuming no immunity to the infectious agent exists within the population.

Bayes's theorem of pre-test probability: concept that observing a phenomenon is unlikely when the occurrence of that phenomenon in the population being observed is low, from eighteenth-century mathematician Thomas Bayes. An important consideration before ordering clinical tests and interpreting results of those tests.

bias: any trend or deviation from the truth in data collection, data analysis, interpretation, and publication which can cause false conclusions.

Bill and Melinda Gates Foundation: private philanthropy group focused on improving LMIC access to vaccines started by Bill Gates, Microsoft founder, and his wife Melinda Gates.

bioavailability: ingredients of a drug remaining after metabolism which retain therapeutic potency at cellular and molecular target levels.

blinding: process by which all participants in a research experiment (subjects, investigators, analysts, and monitors) are prevented from knowing who was randomly assigned to either intervention or control arms of the experiment.

causative agent: any substance (toxin, pathogen, chemical) inducing a disease state.

Centers for Disease Control and Prevention (CDC): federal government agency in the Department of Health and Human Services responsible for national health security and public health initiatives to control disease outbreaks, ensure safety of food and water, prevent leading causes of death, and working with other global agencies to reduce health threats. CDC has some 14,000 staff who work in nearly 170 occupations with field staff in 50 states and more than 50 countries and an annual budget of about $11 billion.

Central Dogma of Molecular Biology: fundamental tenet of modern biology and genetics conceived by Francis Crick stipulating unidirectional flow of genetic information from DNA to RNA to protein, with the possibility of DNA-to-DNA and RNA-to-RNA replication.

clinical test: laboratory or radiology procedure used for medical decision making in individual patients or for public health purposes among whole populations. Healthcare providers interpret data produced by clinical tests to determine if illness is present.

Coalition for Epidemic Preparedness Innovation (CEPI): group formed in 2017 to improve the development, manufacturing, and delivery of vaccines, funded by the European Commission, Wellcome Trust, and Gates Foundation.

codon: three-nucleotide configuration encoding for an amino acid. For example, the codon UGG codes for the amino acid tryptophan.

community seroprevalence studies: public health use of serologic tests to determine widespread immunity developed from infection or vaccination.

comorbid conditions: pre-existing health conditions like diabetes, hypertension, or immunocompromising states predisposing people to becoming afflicted with an ailment or experiencing more serious outcomes.

complementary DNA (cDNA): DNA created by reverse transcribing RNA, a necessary step in DNA amplification during PCR testing.

conferred immunity: immunity achieved through vaccination.

Consolidated Statement of Reporting Trials (CONSORT): standards used by research scientists and journal editors to describe design elements of randomized clinical trials allowing quick assessment of the validity and applicability of clinical trial methods and results.

constitutional federal republic: the governmental and political structure of the United States of America recognizing the Constitution as the "supreme law of the land" for a nation with a centralized federal government performing essential functions on behalf of a republic ruled by the will of the people.

COVID-19 Vaccines Global Access Facility (COVAX): vaccines pillar of the Access to COVID-19 Tools Accelerator (ACT-A) responsible for vaccine distribution to LMICs.

"covidiocy": created from combining "covid" and "idiocy" to describe actions observed during the COVID-19 pandemic that were not in the self-interest of the people engaging in such actions.

cycle threshold value (Ct): number of heating and cooling cycles performed during a PCR assay. If fluorescent probes used in the PCR assay are detectable with a low cycle threshold value, this indicates a high amount of detectable genetic material.

cytochrome P4503A4 (CYP3A4): liver enzyme which metabolizes most medications.

cytoplasmic reproduction: intercellular arena where viral replication occurs.

cytosine: pyrimidine nucleotide found within DNA and RNA molecules that pairs with guanine.

Department of Health and Human Services (DHHS): one of 15 Cabinet departments within the Executive Branch of the federal government containing important healthcare delivery and public health regulatory agencies such as CDC and FDA.

diagnostic test: clinical test used to confirm a suspected diagnosis in someone being evaluated by a healthcare professional.

differential diagnosis: list of possible reasons explaining a patient's complaints generated by health professionals during their encounters with patients based on symptoms, history, and physical exam. Diagnostic tests are often used to confirm the principal diagnosis among all possible reasons contained in the differential diagnosis.

disinformation: misinformation spread intentionally to serve a malicious purpose.

double helix: two parallel DNA strands wound into a right-handed helical shape.

double-stranded: parallel strands of either DNA or RNA molecules.

drug repurposing: therapeutic strategy of using FDA-approved medications for non-FDA-approved purposes when treatment options are limited.

drug-to-drug interactions: adverse effects that may occur when drugs are administered together.

effective reproduction number (Re or Rt): number of people who can be infected by an individual at a specific time. Re (effective reproduction number) or Rt (reproduction number at a specific time) are more precise descriptions of the dynamics of a developing epidemic than R0.

emergency use authorization: FDA administrative procedure used to circumvent normal review activities for approving diagnostics, medications, and vaccines during a public health emergency.

emerging infectious disease (EID): newly appearing infectious diseases caused by previously unknown pathogens or infectious diseases caused by known pathogens within populations not normally infected by these agents.

endoplasmic reticulum: structure of intercellular tubes and sacs through which proteins and other molecules transit as part of their synthetic process.

epitope: specific part of an antigen to which an antibody attaches.

exons: genomic regions which are incorporated into messenger RNA.

Expanded Programme on Immunization (EPI): program launched by World Health Assembly in 1974 to eliminate polio, diphtheria, pertussis, tetanus, and measles among children in LMICs.

false negative: test result indicating the condition being tested for is not present when in fact it is present.

false positive: test result indicating the condition being tested for is present when in fact it is not present.

Federal Food Drug and Cosmetic Act (FFDCA): federal law enacted in 1938 following the catastrophic deaths of over 100 people from ingesting "elixir sulfanilamide" which included a poisonous ingredient. Before FFDCA passage, food processing and drug manufacturing industries were not regulated and had no uniform safety standards or monitoring. FFDCA is a core federal law guiding current work of the FDA.

first pass metabolism: amount of active drug remaining after metabolic actions of the liver.

Food and Drug Administration (FDA): federal government agency in the Department of Health and Human Services overseeing food, medicines, and tobacco products. Among the largest of all federal regulatory agencies, FDA has some 18,000 staff, $6 billion annual budget, and oversees $3 trillion or 20 cents of every dollar spent on products consumed by Americans.

"fourth branch of government": network of regulatory agencies imbued with enough power to be considered an unofficial branch of federal government.

Gavi: private agency focused on vaccine accessibility for low- and middle-income countries founded by the Bill and Melinda Gates Foundation.

genome: entire complement of RNA or DNA genetic material.

gold standard (reference standard): clinical test with very high sensitivity and specificity which almost always correctly identifies a disease as being present or absent.

Golgi body (apparatus, complex): structure of vesicles and folded membranes involved in secretion and intracellular transport.

Great Barrington Declaration: manifesto drafted in Great Barrington, Massachusetts on October 4, 2020 supporting the idea of "focused protection" which encouraged "those who are at minimal risk of death to live their lives normally to build up immunity to the virus through natural infection."

guanine: purine nucleotide found within DNA and RNA molecules that pairs with cytosine.

H1N1 influenza A virus: causative agent of 1918 influenza pandemic continuing to circulate as a virus subtype and remaining a potential pandemic pathogen.

"hammer and dance": alternating cycles of confinement and deconfinement deployed as non-pharmaceutical interventions during infectious disease outbreaks. "hammer" refers to government-imposed restriction of human interactions through lockdowns, quarantines, and social distancing. "Dance" may be either public backlash to these government-imposed measures or palpable relief expressed when people are released from these restrictions.

herd immunity (community immunity; population immunity): indirect protection from infection afforded to susceptible individuals residing in a community where large numbers of others have become immune to that infection through exposure (natural immunity) or vaccination (conferred immunity).

herd immunity threshold: point when forward transmission of a pathogen within a community of susceptible individuals stops because a sufficient percentage have developed immunity.

high-income countries (HICs): World Bank classification of countries by income status (gross national income or GNI per capita) instead of "first world" or "developed country."

host factors: age, sex, socioeconomic status, and other characteristics predisposing people to affliction or causing them to experience more serious clinical stages of illness.

in silico: computer-simulated experimental studies.

in vitro: laboratory-based experimental studies not conducted on animals or humans.

in vivo: laboratory-based experimental studies which are conducted on animals or humans.

incidence: within a population the number of people newly acquiring a certain condition during a specific time period. Incidence measures new cases which develop within a population.

incubation period: amount of time needed for a pathogen to replicate within infected persons before the pathogen is capable of being transmitted from infected to healthy people.

"infodemic": created from combining "information" and "epidemic" defined by WHO as "too much information including false or misleading information in digital or physical environments during a disease outbreak."

innate immune system: first line of immunologic defense against foreign invaders consisting of skin and mucous membranes, gastric acid in the stomach, fever response, and immunologic cells and molecules including macrophages, neutrophils, cytokines, and complement system.

intermediate host: animals which support and amplify zoonotic pathogens facilitating zoonotic spillover of new infections from animals to humans.

International Committee on Taxonomy of Viruses (ICTV): 1966 group responsible for developing, refining, and maintaining a universal virus taxonomy by the Virology Division of the International Union of Microbiological Societies.

Interstate Commerce Commission (ICC): first US federal government regulatory agency established in 1887 to oversee rapid growth in the railroad industry during westward expansion across the continental United States.

introns: genomic regions which are not incorporated within messenger RNA.

John Snow Memorandum: formal response to Great Barrington Declaration named for mid-1800s British physician John Snow who determined how cholera outbreaks occurred in London, the John Snow Memorandum, drafted October 14, 2020, opposed Great Barrington's "focused protection" promoting intentional exposure to SARS-CoV-2 among lower risk healthy people to achieve herd immunity, calling this a "dangerous fallacy unsupported by scientific evidence."

laboratory leak theory: contention that Wuhan Institute of Virology scientists inadvertently or purposefully released modified coronavirus specimens into the surrounding community causing the first cases of COVID-19.

lateral flow immunoassay (LFIA): paper-based chromatography testing method using capillary forces to carry antigenic contents of a specimen solution along a series of absorbing pads imbued with reactive antibodies and signaling presence of the antigenic content as a "positive test."

logical fallacies: accidental or manipulative arguments based on incorrect reasoning leading to erroneous conclusions.

low-and middle-income countries (LMICs): World Bank classification of countries by income status (gross national income or GNI per capita) instead of "third world" or "developing country."

main protease (Mpro; 3CL-protease): SARS-CoV-2 viral protease enzyme which cleaves elongated chains of inactivated polyproteins pp1a and pp1ab into active proteins necessary for continuing viral replication.

major histocompatibility complexes (MHCs): structures within host cells which detect foreign proteins like viruses, package these foreign substances into fragments, and push these packaged fragments to external surfaces of infected host cell membranes as a signal to the immune system of foreign invasion.

medical countermeasures (MCMs): medicines and medical supplies used to diagnose, prevent, protect from, or treat conditions associated with chemical, biological, radiological, or nuclear (CBRN) threats or emerging infectious diseases. Examples include diagnostic tests, therapeutic agents, and vaccines.

membranous envelope: outer shell of a virus which encases its genetic material.

misinformation: health information which is false, inaccurate, or misleading according to best available evidence at the time.

monoclonal antibodies: antibodies created by single clone of plasma cells able to target specific antigens and used as "passive immunization" during early treatment of an infectious disease.

multisystem inflammatory syndrome in children (MIS-C): rare condition reported in children with COVID-19 characterized by fever, hypotension, and multiple organ involvement.

natural immunity: immunity achieved through infection alone.

negative predictive value (NPV): when interpreting a negative test result, proportion who truly do not have the condition and truly have a negative test result among all people with a negative test result, some of whom have truly negative test results and some having false negative results.

non-pharmaceutical interventions: activities which prevent transmission of pathogens during infectious disease outbreaks, such as social distancing, facemask wearing, and quarantines.

notice and comment period: APA provision allowing direct public participation in how regulations are created when a "notice" or announcement about a proposed regulation being considered is published in the *Federal Register* allowing a "comment" period of 90 days during which written reactions from the public about the proposed regulation can be made.

nucleic acid vaccines: use the pathogen's own DNA or RNA to encode important proteins causing infection to instruct human cells to create the antigenic components against which an immune response is generated. SARS-CoV-2 vaccines using messenger RNA technology are examples of nucleic acid vaccines.

nucleotides: nuclei acid molecules consisting of a ribose sugar (in RNA molecules) or deoxyribose sugar (in DNA molecules) with a phosphate group and nitrogenous base.

open reading frame: portion of RNA or DNA molecule which contains no stop codons and allows unimpeded translation.

organelles: membrane-enclosed structures within eukaryotic cells such as endoplasmic reticulum, Golgi bodies, mitochondria, and nuclei.

pathogen: a disease-causing organism.

pathogen-associated molecular pattern (PAMP): molecular features unique to pathogens which are recognized as being foreign by the human immune system.

pathophysiology: disordered physiological processes which give rise to a disease state.

peer-reviewed journals: periodicals such as the *New England Journal of Medicine* or the *British Medical Journal* in which findings from experiments conducted by researchers are written up as articles and published. This allows experimental findings to be scrutinized or "peer-reviewed" by others in the field of study in which the experiments were conducted. Peer reviewers serve as informal judges rendering opinions about the veracity of published experimental findings, expressing these opinions in "letters to the editor" of the journals in which results appeared.

pharmacodynamics: branch of pharmacology which studies relationships between drug concentration at site of action and resulting therapeutic or adverse effects from this interaction.

pharmacokinetics: branch of pharmacology which studies the time course of movement ("kinetics") of a drug through phases of absorption, distribution, metabolism, and excretion.

Phylogenetic Assignment of Named Global Outbreak (PANGO): computational tool and nomenclature scheme developed in April 2020 to assign the most likely phylogenetic lineage to a given SARS-CoV-2 genome sequence.

placebo: inert substance similar in appearance to the therapeutic agent being tested.

positive predictive value: when interpreting a positive test result, proportion who truly have the condition and also have a positive test result among all people having a positive test result, some of whom have true positive test results and some having false positive test results.

positive sense: nucleotide sequence which directly corresponds to the nucleotide sequence of messenger RNA ready to be translated into protein.

post-acute sequelae of COVID-19 (PASC; "long COVID"): constellation of lingering symptoms in certain COVID-19 patients which is still being defined.

preprint server: publishing website where findings from scientific experiments can be uploaded without being peer-reviewed, thereby circumventing any judgment by peer reviewers about the veracity of results claimed by its authors. The biomedical preprint server *bioRxiv* maintained by Cold Spring Harbor Laboratory is an example. Some articles appearing on preprint servers have been or will be submitted to peer-reviewed journals for final publication.

prevalence: within a population the specific number of people newly acquiring or previously having a certain condition during a specific time period. Prevalence measures all cases, both new cases and existing cases, which are present within a population.

protease: enzyme which can cleave polypeptides strands into small peptides.

protease inhibitor: therapeutic agent which inhibits actions of proteases.

Public Health Service Act (PHSA): federal law enacted in 1944 to harness public health activities of multiple federal agencies and departments, the cornerstone of work conducted by CDC and other federal entities.

pulmonary alveoli: small sacs enmeshed in capillaries found at the furthest point in the respiratory tract where oxygen and carbon dioxide gas exchange occurs.

pulmonary surfactant: phospholipid and protein complex which facilitates gas exchange within pulmonary alveoli and is formed by type II alveolar epithelial cells.

quantitative real-time polymerase chain reaction test (qRT-PCR or PCR): laboratory technique which detects genes of interest from DNA samples using a heat-resistant DNA polymerase enzyme to produce multiple copies of genetic material fragments.

randomization: process used during a clinical trial which randomly assigns subjects to intervention and control arms, distributing these subjects using a computer algorithm to ensure both arms contain persons sharing similar characteristics such as age, gender, or race.

randomized clinical trial (RCT): experiment which randomly assigns subjects to an intervention arm containing the actual medication, device, or procedure being studied and a control arm containing an innocuous placebo or inert substance. Bias is reduced when experiment designers, investigators, and subjects themselves are prevented from knowing whether subjects have been relegated to intervention or control arms, a shielding process called "blinding." RCTs are considered to yield the strongest objective evidence about the veracity of an intervention but can be costly, time-consuming, and have ethical concerns associated with their performance.

renin-angiotensin-aldosterone system (RAAS): critical physiologic mechanism maintaining fluid balance and blood pressure control.

replication-incompetent: genetic material which may continue to be detected by the PCR amplification process but does not cause an actual infection.

ribosome: intercellular structure comprising RNA and protein playing a principal role in protein synthesis.

RNA-dependent RNA polymerase (RdRp): produced by all RNA viruses, RdRp plays crucially important roles in RNA-to-RNA information transfer by transcribing a copy of the entire RNA genome which will be encapsulated within newly forming viruses (called "copy genomic RNA") and transcribing fragmented copies of mRNA used as translation templates to make other proteins (called "subgenomic RNA").

rulemaking: process for creating detailed regulations established in APA incorporating a "notice and comment" activity allowing public participation in creating regulations.

S-gene target failure (SGTF): early PCR test indicator of possible mutation(s) in the S-gene coding sequence for SARS-CoV-2 spike protein.

scientific method: a process which explains natural phenomena through a series of activities: developing a hypothesis about the observed phenomenon; designing experiments to prove or disprove the hypothesis; interpreting observations from these experiments; and using these observations to refine the original hypothesis or to develop new hypotheses.

screening test: clinical test detecting presence of an illness or condition in a healthy person.

sensitivity: ability of a test to indicate someone truly has a condition.

serologic antibody test: blood tests detecting immunoglobulins created during infection.

single-stranded: one strand of either DNA or RNA molecules.

specificity: ability of a test to indicate someone truly does not have a condition.

spike glycoprotein: antigen target of SARS-CoV-2.

start codon: mRNA sequence (AUG) coding for methionine, signaling start of protein synthesis.

stop codon: three mRNA sequences (UAA, UAG, UGA) signaling end of protein synthesis.

subunit vaccines: contain parts of whole pathogen, such as portions of its cell membrane or other protein components. These non-infectious portions are the antigenic components against which an immune response is generated. Meningococcal vaccines are examples of subunit vaccines.

Surgeon General's Advisories: official reports issued by Surgeons General about serious public health matters beginning with the first Advisory in 1964 entitled "Smoking and Health: Report of the Advisory Committee to the Surgeon General" and the most recent July 2021 Advisory "Confronting Health Misinformation: the U.S. Surgeon General's Advisory on Building a Healthy Information Environment."

susceptible-infected-recovered (SIR) forecasting model: mathematical approach for predicting possible scenarios of pathogen transmission, infection cases, and recovery at the start of a new infectious disease outbreak when real-world data are not yet available.

symptom-based approach: CDC recommended way to discontinue home isolation for SARS-CoV-2 positive persons based on having no further symptoms of COVID-19 illness.

symptoms: self-perceived discomfort such as fever, body aches, cough, tiredness, or stomach upset which may indicate a beginning illness.

Taq DNA polymerase: heat-resistant polymerase enzyme harvested from heat-tolerant *Thermus aquatics* ("Taq" is an abbreviation combing the "T" of *Thermus* and the "aq" of *aquatics*) used during the polymerase chain reaction process. Heat-resistant polymerase enzyme is necessary as PCR accomplishes its nucleic acid amplification through a series of heating and cooling cycles.

ten-word definition of life: "life is a self-sustaining chemical system capable of Darwinian evolution" coined by NASA as a means to evaluate extra-terrestrial living systems.

therapeutic agent: chemical or biological substance which prevents, alters, or reverses a disease process.

thymine: pyrimidine nucleotide found only in DNA molecules that pairs with adenine.

time-based approach: CDC recommended way to discontinue home isolation for SARS-CoV-2 positive persons based on number of days in isolation during a COVID-19 illness.

transcription: process of copying genetic information on DNA strands into messenger RNA.

translation: protein synthesis enabled by coding sequence within messenger RNA.

transmembrane protease serine 2 (TMPRSS2): host cell membrane protease that assists SARS-CoV-2 entry into host cells by priming actions.

tropism: preferential tendency a virus has for a specific cell or tissue as its infection target.

true negative: test result which means the condition being tested for is truly not present.

true positive: test result which means the condition being tested for is truly present.

turnaround time (TAT): amount of time needed to produce a clinical test result.

Type I alveolar epithelial cells (AEC1): specialized epithelial cells which maintain the structural integrity of the alveolus-capillary interface.

Type II alveolar epithelial cells (AEC2): specialized epithelial cells which produce pulmonary surfactant, replenish AEC1 cells, and participate in immune defense of the alveolar space.

uracil: pyrimidine nucleotide found only in RNA molecules that pairs with adenine.

Vaccines and Related Biological Products Advisory Committee (VRBPAC): FDA advisory group of government and private sector experts which reviews the scientific merits of clinical trial data from vaccine and biological product research, developing recommendations which are sent to the FDA Commissioner for final approval.

variolation: historical method of using pus extracted from vesicles to inoculate healthy subjects, used most notably by British physician Edward Jenner during his smallpox investigations.

viral load: amount of virus in an infected person able to be transmitted to healthy people.

WHO Greek letter labels: naming convention recommended by WHO in May 2021 as a non-judgmental way to discuss emerging SARS-CoV-2 variants without blaming countries whose genomic surveillance systems first detected the emerging variant. For instance, the B.1.1.7 variant first noted by UK genomic surveillance in September 2020 is called the "Alpha variant."

whole-pathogen vaccines: contain the infecting pathogen itself, which has been weakened (called "live-attenuated virus vaccine") or killed (called "killed of inactivated vaccine") by physical or chemical means so as not to cause actual infection when injected. The mumps, measles, rubella vaccine (MMR) is an example of a live-attenuated virus vaccine.

Wuhan Institute of Virology (WIV): Biosafety Level 4 laboratory located in the city of Wuhan in the Chinese province of Hubei considered the originating source for SARS-CoV-2 by adherents of the laboratory leak theory.

zoonotic pathogens: most often viruses from vertebrate animals that may "jump species" and cause new infectious diseases in humans.

zoonotic spillover theory: emerging infection disease caused by an animal-based (zoonotic) pathogen "jumping" from animal to human.

zugzwang: chess dilemma in which a game movement must be made even if the move is detrimental to the player, from the German word "zug" (move) "zwang" (obligation), used as a metaphor for the conflicting decisions faced by world leaders during the COVID-19 pandemic.

Index

Note: The page preference with letter n represents note numbers.

Printed in the United States
by Baker & Taylor Publisher Services

Printed in the United States
by Baker & Taylor Publisher Services